Delmar's Test Preparation

Automobile Test

Engine Repair (Test A1)

Technical Advisor
Boyce H. Dwiggins

Delmar Publishers

an International Thomson Publishing company I(T)P®

Albany • Bonn • Boston • Cincinnati • Detroit • London • Madrid
Melbourne • Mexico City • New York • Pacific Grove • Paris • San Francisco
Singapore • Tokyo • Toronto • Washington

NOTICE TO THE READER

Cover Design: Paul Roseneck

Delmar Staff:
Publisher: Alar Elken
Acquisitions Editor: Vernon Anthony
Editorial Assistant: Betsy Hough
Marketing Manager: Mona Caron

COPYRIGHT © 1999
By Delmar Publishers
an International Thomson Publishing company I(T)P®

The ITP logo is a trademark under license
Printed in the United States of America

Online Services

Delmar Online
To access a wide variety of Delmar products and services on the World Wide Web, point your browser to:
http://www.delmar.com
or email: info@delmar.com

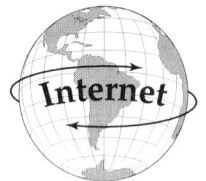

A service of I(T)P®

For more information contact:

Delmar Publishers
3 Columbia Circle, Box 15015
Albany, New York 12212-5015

International Thomson Publishing Europe
Berkshire House
168-173 High Holborn
London, WC1V7AA
United Kingdom

Nelson ITP, Australia
102 Dodds Street
South Melbourne,
Victoria, 3205 Australia

Nelson Canada
1120 Birchmont Road
Scarborough, Ontario
M1K 5G4, Canada

International Thomson Publishing France
Tour Maine-Montparnasse
33 Avenue du Maine
75755 Paris Cedex 15, France

International Thomson Editores
Seneca 53
Colonia Polanco
11560 Mexico D. F. Mexico

International Thomson Publishing GmbH
Königswinterer Strasse 418
53227 Bonn
Germany

International Thomson Publishing Asia
60 Albert Street
#15-01 Albert Complex
Singapore 189969

International Thomson Publishing Japan
Hirakawa-cho Kyowa Building, 3F
2-2-1 Hirakawa-cho, Chiyoda-ku,
Tokyo 102, Japan

ITE Spain/ Paraninfo
Calle Magallanes, 25
28015-Madrid, Spain

3 4 5 6 7 8 9 10   XXX   03 02 01 00 99

ISBN 0-7668-0549-2

Contents

Section 3 Are You Sure You're Ready for A1 Test?

Section 4 An Overview of the System

Section 5 Sample Test for Practice

Section 6 Additional Test Questions for Practice

Section 7 Appendices

Preface

This book is just one of a comprehensive series designed to prepare technicians to take and pass every ASE test. Delmar's series covers all of the Automotive tests A1 through A8 as well as Advanced Engine Performance L2 and Parts Specialist P2. The series also covers the five Collision Repair tests and the eight Medium/Heavy Duty truck test.

Before any book in this series was written, Delmar staff met with and surveyed technicians and shop owners who have taken ASE tests and have used other preparatory materials. We found that they wanted, first and foremost, *lots* of practice tests and questions. Each book in our series contains a general knowledge pretest, a sample test, and additional practice questions. You will be hard-pressed to find a test prep book with more questions for you to practice with. We have worked hard to ensure that these questions match the ASE style in types of questions, quantities, and level of difficulty.

Technicians also told us that they wanted to understand the ASE test and to have practical information about what they should expect. We have provided that as well, including a history of ASE and a section devoted to helping the technician "Take and Pass Every ASE Test" with case studies, test-taking strategies, and test formats.

Finally, techs wanted refresher information and reference. Each of our books includes an overview section that is referenced to the task list. The complete task lists for each test appear in each book for the user's reference. There is also a complete glossary of terms for each booklet.

So whether you're looking for a sample test and a few extra questions to practice with or a complete introduction to ASE testing, with support for preparing thoroughly, this book series is an excellent answer.

We hope you benefit from this book and that you pass every ASE test you take!

Your comments, both positive and negative, are certainly encouraged! Please contact us at:

Automotive Editor
Delmar Publishers
3 Columbia Circle
Box 15015
Albany, NY 12212-5015

1 The History of ASE

History

Originally known as The National Institute for Automotive Service Excellence (NIASE), today's ASE was founded in 1972 as a non-profit, independent entity dedicated to improving the quality of automotive service and repair through the voluntary testing and certification of automotive technicians. Until that time, consumers had no way of distinguishing between competent and incompetent automotive mechanics. In the mid-1960s and early 1970s, efforts were made by several automotive industry affiliated associations to respond to this need. Though the associations were non-profit, many regarded certification test fees merely as a means of raising additional operating capital. Also, some associations, having a vested interest, produced test scores heavily weighted in the favor of its members.

NIASE

From these efforts a new independent, non-profit association, the National Institute for Automotive Service Excellence (NIASE), was established much to the credit of two educators, George R. Kinsler, Director of Program Development for the Wisconsin Board of Vocational and Adult Education in Madison, WI, and Myron H. Appel, Division Chairman at Cypress College in Cypress, CA.

Early efforts were to encourage voluntary certification in four general areas:

TEST AREA	TITLES
I. Engine	Engines, Engine Tune-Up, Block Assembly, Cooling and Lube Systems, Induction, Ignition, and Exhaust
II. Transmission	Manual Transmissions, Drive Line and Rear Axles, and Automatic Transmissions
III. Brakes and Suspension	Brakes, Steering, Suspension, and Wheels
IV. Electrical/Air Conditioning	Body/Chassis, Electrical Systems, Heating, and Air Conditioning

In early NIASE tests, Mechanic A, Mechanic B type questions were used. Over the years the trend has not changed, but in mid-1984 the term was changed to Technician A, Technician B to better emphasize sophistication of the skills needed to perform successfully in the modern motor vehicle industry. In certain tests the term used is Estimator A/B, Painter A/B, or Parts Specialist A/B. At about that same time, the logo was changed from "The Gear" to "The Blue Seal," and the organization adopted the acronym ASE for Automotive Service Excellence.

Since those early beginnings, several other related trades have been added. ASE now administers a comprehensive series of certification exams for automotive and light truck repair technicians, medium and heavy truck repair technicians, alternate fuels technicians, engine machinists, collision repair technicians, school bus repair technicians, and parts specialists.

The Series and Individual Tests

- Automotive and Light Truck Technician; consisting of: Engine Repair - Automatic Transmission/Transaxle - Manual Drive Train and Axles - Suspension and Steering - Brakes - Electrical/Electronic Systems - Heating and Air Conditioning - Engine Performance
- Medium and Heavy Truck Technician; consisting of: Gasoline Engines - Diesel Engines - Drive Train - Brakes - Suspension and Steering - Electrical/Electronic Systems - HVAC - Preventive Maintenance Inspection
- Alternate Fuels Technician; consisting of: Compressed Natural Gas Light Vehicles
- Advanced Series; consisting of: Automobile Advanced Engine Performance and Advanced Diesel Engine Electronic Diesel Engine Specialty
- Collision Repair Technician; consisting of: Painting and Refinishing - Non-Structural Analysis and Damage Repair - Structural Analysis and Damage Repair - Mechanical and Electrical Components - Damage Analysis and Estimating
- Engine Machinist Technician; consisting of: Cylinder Head Specialist - Cylinder Block Specialist - Assembly Specialist
- School Bus Repair Technician; consisting of: Body Systems and Special Equipment - Drive Train - Brakes - Suspension and Steering - Electrical/Electronic Systems - Heating and Air Conditioning
- Parts Specialist; consisting of: Automobile Parts Specialist - Medium/Heavy Truck Parts Specialist

A Brief Chronology

1970–1971	Original questions were prepared by a group of forty auto mechanics teachers from public secondary schools, technical institutes, community colleges, and private vocational schools. These questions were then professionally edited by testing specialists at Educational Testing Service (ETS) at Princeton, New Jersey, and thoroughly reviewed by training specialists associated with domestic and import automotive companies.
1971	July: About eight hundred mechanics tried out the original test questions at experimental test administrations.
1972	November and December: Initial NIASE tests administered at 163 test centers. The original automotive test series consisted of four tests containing eighty questions each. Three hours were allotted for each test. Those who passed all four tests were designated Certified General Auto Mechanic (GAM).
1973	April and May: Test 4 was increased to 120 questions. Time was extended to four hours for this test. There were now 182 test centers. Shoulder patch insignias were made available.

	November: Automotive series expanded to five tests. Heavy-Duty Truck series of six tests introduced.
1974	November: Automatic Transmission (Light Repair) test modified. Name changed to Automatic Transmission.
1975	May: Collision Repair series of two tests is introduced.
1978	May: Automotive recertification testing is introduced.
1979	May: Heavy-Duty Truck recertification testing is introduced.
1980	May: Collision Repair recertification testing is introduced.
1982	May: Test administration providers switched from Educational Testing Service (ETS) to American College Testing (ACT). Name of Automobile Engine Tune-Up test changed to Engine Performance test.
1984	May: New logo was introduced. ASE's "The Blue Seal" replaced NIASE's "The Gear." All reference to Mechanic A, Mechanic B was changed to Technician A, Technician B.
1990	November: The first of the Engine Machinist test series was introduced.
1991	May: The second test of the Engine Machinist test series was introduced. November: The third and final Engine Machinist test was introduced.
1992	May: Name of Heavy-Duty Truck Test series changed to Medium/Heavy Truck Test series.
1993	May: Automotive Parts Specialist test introduced. Collision Repair expanded to six tests. Light Vehicle Compressed Natural Gas test introduced. Limited testing begins in English-speaking provinces of Canada.
1994	May: Advanced Engine Performance Specialist test introduced.
1996	May: First three tests for School Bus Technician test series introduced. November: A Collision Repair test is added.
1997	May: A Medium/Heavy Truck test is added.
1998	May: A diesel advanced engine test is introduced: Electronic Diesel Engine Diagnosis Specialist. A test is added to the School Bus test series.

By the Numbers

Following are the approximate number of ASE technicians currently certified by category. The numbers may vary from time to time but are reasonably accurate for any given period. More accurate data may be obtained from ASE, which provides updates twice each year, in May and November after the Spring and Fall test series.

There are more than 338,000 Automotive Technicians with over 87,000 at Master Technician (MA) status. There are 47,000 Truck Technicians with over 19,000 at Master Technician (MT) status. There are 46,000 Collision Repair/Refinish Technicians with 7,300 at Master Technician (MB) status. There are 1,200 Estimators. There are 6,700 Engine Machinists with over 2,800 at Master Machinist Technician (MM) status. There are also 28,500 Automobile Advanced Engine Performance Technicians and over 2,700 School Bus Technicians for a combined total of more than 403,000 Repair Technicians. To this number, add over 22,000 Automobile Parts Specialists, and over 2,000 Truck Parts Specialists for a combined total of over 24,000 parts specialists.

There are over 6,400 ASE Technicians holding both Master Automotive Technician and Master Truck Technician status, of which 350 also hold Master Body Repair status. Almost 200 of these Master Technicians also hold Master Machinist status and five Technicians are certified in all ASE specialty areas.

Almost half of ASE certified technicians work in new vehicle dealerships (45.3%). The next greatest number work in independent garages with 19.8%. Next is tire dealerships with 9%, service stations at 6.3%, fleet shops at 5.7%, franchised volume retailers at 5.4%, paint and body shops at 4.3%, and specialty shops at 3.9%.

Of over 400,000 automotive technicians on ASE's certification rosters, almost 2,000 are female. The number of female technicians is increasing at a rate of about 20% each year. Women's increasing interest in automotive mechanics is further evidenced by the fact that, according to the National Automobile Dealers Association (NADA), they influence 80% of the decisions of the purchase of a new automobile and represent 50% of all new car purchasers. Also, it is interesting to note that 65% of all repair and maintenance service customers are female.

The typical ASE certified technician is 36.5 years of age, is computer literate, deciphers a half-million pages of technical manuals, spends one hundred hours per year in training, holds four ASE certificates, and spends about $27,000 for tools and equipment. Twenty-seven percent of today's skilled ASE certified technicians attended college, many having earned an Associate of Science degree in Automotive Technology.

ASE

ASE's mission is to improve the quality of vehicle repair and service in the United States through the testing and certification of automotive repair technicians. Prospective candidates register for and take one or more of ASE's thirty-three exams. The tests are grouped into specialties for automobile, medium/heavy truck, school bus, and collision repair technicians as well as engine machinists, alternate fuels technicians, and parts specialists.

Upon passing at least one exam and providing proof of two years of related work experience, the technician becomes ASE certified. A technician who passes a series of exams earns ASE Master Technician status. An automobile technician, for example, must pass eight exams for this recognition.

The tests, conducted twice a year at over seven hundred locations around the country, are administered by American College Testing (ACT). They stress real-world diagnostic and repair problems. Though a good knowledge of theory is helpful to the technician in answering many of the questions, there are no questions specifically on theory. Certification is valid for five years. To retain certification, the technician must be retested to renew his or her certificate.

The automotive consumer benefits because ASE certification is a valuable yardstick by which to measure the knowledge and skills of individual technicians, as well as their commitment to their chosen profession. It is also a tribute to the repair facility employing ASE certified technicians. ASE certified technicians are permitted to wear blue and white ASE shoulder insignia, referred to as the "Blue Seal of Excellence," and carry credentials listing their areas of expertise. Often employers display their technicians' credentials in the customer waiting area. Customers look for facilities that display ASE's Blue Seal of Excellence logo on outdoor signs, in the customer waiting area, in the telephone book (Yellow Pages), and in newspaper advertisements.

The tests stress repair knowledge and skill. All test takers are issued a score report. In order to earn ASE certification, a technician must pass one or more of the exams and present proof of two years of relevant hands-on work experience. ASE certifications are valid for five years, after which time technicians must retest in order to keep up with changing technology and to remain in the ASE program. A nominal registration and test fee is charged.

To become part of the team that wear ASE's Blue Seal of Excellence®, please contact:

National Institute for Automotive Service Excellence
13505 Dulles Technology Drive
Herndon, VA 20171-3421

2 Take and Pass Every ASE Test

ASE Testing

Participating in an Automotive Service Excellence (ASE) voluntary certification program gives you a chance to show your customers that you have the "know-how" needed to work on today's modern vehicles. The ASE certification tests allow you to compare your skills and knowledge to the automotive service industry's standards for each specialty area.

If you are the "average" automotive technician taking this test, you are in your mid-thirties and have not attended school for about fifteen years. That means you probably have not taken a test in many years. Some of you, on the other hand, have attended college or taken postsecondary education courses and may be more familiar with taking tests and with test-taking strategies. There is, however, a difference in the ASE test you are preparing to take and the educational tests you may be accustomed to.

Who Writes the Questions?

The questions on an educational test are generally written, administered, and graded by an educator who may have little or no practical hands-on experience in the test area. The questions on all ASE tests are written by service industry experts familiar with all aspects of the subject area. ASE questions are entirely job-related and designed to test the skills that you need to know on the job.

The questions originate in an ASE "item-writing" workshop where service representatives from domestic and import automobile manufacturers, parts and equipment manufacturers, and vocational educators meet in a workshop setting to share their ideas and translate them into test questions. Each test question written by these experts is reviewed by all of the members of the group. The questions deal with the practical problems of diagnosis and repair that are experienced by technicians in their day-to-day hands-on work experiences.

All of the questions are pretested and quality-checked in a nonscoring section of tests by a national sample of certifying technicians. The questions that meet ASE's high standards of accuracy and quality are then included in the scoring sections of future tests. Those questions that do not pass ASE's stringent test are sent back to the workshop or are discarded. ASE's tests are monitored by an independent proctor and are administered and machine-scored by an independent provider, American College Testing (ACT). All ASE tests have a three-year revision cycle.

Testing

If you think about it, we are actually tested on about everything we do. As infants, we were tested to see when we could turn over and crawl, later when we could walk or talk. As adolescents, we were tested to determine how well we learned the material presented in school and in how we demonstrated our accomplishments on the athletic field. As working adults, we are tested by our supervisors on how well we have completed an assignment or project. As nonworking adults, we are tested by our family on everyday activities, such as housekeeping or preparing a meal. Testing, then, is one of those facts of life that begins in the cradle and follows us to the grave.

Testing is an important fact of life that helps us to determine how well we have learned our trade. Also, tests often help us to determine what opportunities will be available to us in the future. To become ASE certified, we are required to take a test in every subject in which we wish to be recognized.

Be Test-Wise

In spite of the widespread use of tests, most technicians are not very test-wise. An ability to take tests and score well is a skill that must be acquired. Without this knowledge, the most intelligent and prepared technician may not do well on a test.

We will discuss some of the basic procedures necessary to follow in order to become a test-wise technician. Assume, if you will, that you have done the necessary study and preparation to score well on the ASE test.

Different approaches should be used for taking different types of tests. The different basic types of tests include: essay, objective, multiple choice, fill in the blank, true-false, problem solving, and open book. All ASE tests are of the four-part multiple-choice type.

Before discussing the multiple-choice type test questions, however, there are a few basic principles that should be followed before taking any test.

Before the Test

Do not arrive late. Always arrive well before your test is scheduled to begin. Allow ample time for the unexpected, such as traffic problems, so you will arrive on time and avoid the unnecessary anxiety of being late.

Always be certain to have plenty of supplies with you. For an ASE test, three or four sharpened soft lead (#2) pencils, a pocket pencil sharpener, erasers, and a watch are all that are required.

Do not listen to pretest chatter. When you arrive early, you may hear other technicians testing each other on various topics or making their best guess as to the probable test questions. At this time, it is too late to add to your knowledge. Also the rhetoric may only confuse you. If you find it bothersome, take a walk outside the test room to relax and loosen up.

Read and listen to all instructions. It is important to read and listen to the instructions. Make certain that you know what is expected of you. Listen carefully to verbal instructions and pay particular attention to any written instructions on the test paper. Do not dive into answering questions only to find out that you have answered the wrong question by not following instructions carefully. It is difficult to make a high score on a test if you answer the wrong questions.

These basic principles have been violated in most every test ever given. Try to remember them. They are essential for success.

Objective Tests

A test is called an objective test if the same standards and conditions apply to everyone taking the test and there is only one correct answer to each question. Objective tests primarily measure your ability to recall information. A well-designed objective test can also test your ability to understand, analyze, interpret, and apply your knowledge. Objective tests include true-false, multiple choice, fill in the blank, and matching questions.

Objective questions, not generally encountered in a classroom setting, are frequently used in standardized examinations. Objective tests are easy to grade and also reduce the amount of paperwork necessary to administer. The objective tests are used in entry-level programs or when very large numbers are being tested. ASE's tests consist exclusively of four-part multiple-choice objective questions in all of their tests.

Taking an Objective Test

The principles of taking an objective test are somewhat different from those used in other types of tests. You should first quickly look over the test to determine the number of questions, but do not try to read through all of the questions. In an ASE test, there are usually between forty and eighty questions, depending on the subject matter. Read through each question before marking your answer. Answer the questions in the order they appear on the test. Leave the questions blank that you are not sure of and move on to the next question. You can return to those unanswered questions after you have finished the others. They may be easier to answer at a later time after your mind has had additional time to consider them on a subconscious level. In addition, you might find information in other questions that will help you to answer some of them.

Do not be obsessed by the apparent pattern of responses. For example, do not be influenced by a pattern like **d**, **c**, **b**, **a**, **d**, **c**, **b**, **a** on an ASE test.

There is also a lot of folk wisdom about taking objective tests. For example, there are those who would advise you to avoid response options that use certain words such as *all*, *none*, *always*, *never*, *must*, and *only*, to name a few. This, they claim, is because nothing in life is exclusive. They would advise you to choose response options that use words that allow for some exception, such as *sometimes*, *frequently*, *rarely*, *often*, *usually*, *seldom*, and *normally*. They would also advise you to avoid the first and last option (A and D) because test writers, they feel, are more comfortable if they put the correct answer in the middle (B and C) of the choices. Another recommendation often offered is to select the option that is either shorter or longer than the other three choices because it is more likely to be correct. Some would advise you to never change an answer since your first intuition is usually correct.

Although there may be a grain of truth in this folk wisdom, ASE test writers try to avoid them and so should you. There are just as many **A** answers as there are **B** answers, just as many **D** answers as **C** answers. As a matter of fact, ASE tries to balance the answers at about 25 percent per choice **A**, **B**, **C**, and **D**. There is no intention to use "tricky" words, such as outlined above. Put no credence in the opposing words "sometimes" and "never," for example. When used in an ASE type question, one or both may be correct; one or both may be incorrect.

There are some special principles to observe on multiple-choice tests. These tests are sometimes challenging because there are often several choices that may seem possible, and it may be difficult to decide on the correct choice. The best strategy, in this case, is to first determine the correct answer before looking at the options. If you see the answer you decided on, you should still examine the options to make sure that none seem more correct than yours. If you do not know or are not sure of the answer, read each option very carefully and try to eliminate those options that you know to be wrong. That way, you can often arrive at the correct choice through a process of elimination.

If you have gone through all of the test and you still do not know the answer to some of the questions, then guess. Yes, guess. You then have at least a 25 percent chance of being correct. If you leave the question blank, you have no chance. In ASE tests, there is no penalty for being wrong. As the late President Franklin D. Roosevelt once advised a group of students, "It is common sense to take a method and try it. If it fails, admit it frankly and try another. But above all, try something."

During the Test

Mark your bubble sheet clearly and accurately. One of the biggest problems an adult faces in test-taking, it seems, is in placing an answer in the correct spot on a bubble sheet. Make certain that you mark your answer for, say, question 21, in the space on the bubble sheet designated for the answer for question 21. A correct response in the wrong bubble will probably be wrong. Remember, the answer sheet is machine scored and can only "read" what you have bubbled in. Also, do not bubble in two answers for the same question. For example, if you feel the answer to a particular question is **A** but think it may be **C**, do not bubble in both choices. Even if either **A** or **C** is correct, a double answer will score as an incorrect answer. It's better to take a chance with your best guess.

Review Your Answers

If you finish answering all of the questions on a test ahead of time, go back and review the answers of those questions that you were not sure of. You can often catch careless errors by using the remaining time to review your answers.

Don't Be Distracted

At practically every test, some technicians will invariably finish ahead of time and turn their papers in long before the final call. Do not let them distract or intimidate you. Either they knew too little and could not finish the test, or they were very self-confident and thought they knew it all. Perhaps they were trying to impress the proctor or other technicians about how much they know. Often you may hear them later talking about the information they knew all the while but forgot to respond on their answer sheet.

Use Your Time Wisely

It is not wise to use less than the total amount of time that you are allotted for a test. If there are any doubts, take the time for review. Any product can usually be made better with some additional effort. A test is no exception. It is not necessary to turn in your test paper until you are told to do so.

Don't Cheat

Some technicians may try to use a "crib sheet" during a test. Others may attempt to read answers from another technician's paper. If you do that, you are unquestionably assuming that someone else has a correct answer. You probably know as much, maybe more, than anyone else in the test room. Trust yourself. If you're still not convinced, think of the consequences of being caught. Cheating is foolish. If you are caught, you have failed the test.

Be Confident

The first and foremost principle in taking a test is that you need to know what you are doing, to be test-wise. It will now be presumed that you are a test-wise technician and are now ready for some of the more obscure aspects of test-taking.

An ASE-style test requires that you use the information and knowledge at your command to solve a problem. This generally requires a combination of information similar to the way you approach problems in the real world. Most problems, it seems, typically do not fall into neat textbook cases. New problems are often difficult to handle, whether they are encountered inside or outside the test room.

An ASE test also requires that you apply methods taught in class as well as those learned on the job to solve problems. These methods are akin to a well-equipped tool box in the hands of a skilled technician. You have to know what tools to use in a particular situation, and you must also know how to use them. In an ASE test, you will need to be able to demonstrate that you are familiar with and know how to use the tools.

You should begin a test with a completely open mind. At times, however, you may have to move out of your normal way of thinking and be creative to arrive at a correct answer. If you have diligently studied for at least one week prior to the test, you have bombarded your mind with a wide assortment of information. Your mind will be working with this information on a subconscious level, exploring the interrelationships among various facts, principles, and ideas. This prior preparation should put you in a creative mood for the test.

In order to reach your full potential, you should begin a test with the proper mental attitude and a high degree of self-confidence. You should think of a test as an opportunity to document how much you know about the various tasks in your chosen profession. If you have been diligently studying the subject matter, you will be able to take your test in serenity because your mind will be well organized. If you are confident, you are more likely to do well because you have the proper mental attitude. If, on the other hand, your confidence is low, you are bound to do poorly. It is a self-fulfilling prophecy.

Perhaps you have heard athletic coaches talk about the importance of confidence when competing in sports. Mental confidence helps an athlete to perform at the highest level and gain an advantage over competitors. Taking a test is much like an athletic event. You are competing against yourself, in a certain sense, because you will be trying to approach perfection in determining your answers. As in any competition, you should aim your sights high and be confident that you can reach the apex.

Anxiety and Fear

Many technicians experience anxiety and fear at the very thought of taking a test. Many worry, become nervous, and even become ill at test time because of the fear of failure. Many often worry about the criticism and ridicule that may come from their employer, relatives, and peers. Some worry about taking a test because they feel that the stakes are very high. Those who spent a great amount of time studying may feel they must get a high grade to justify their efforts. The thought of not doing well can result in unnecessary worry. They become so worried, in fact, that their reasoning and thinking ability is impaired, actually bringing about the problem they wanted to avoid.

The fear of failure should not be confused with the desire for success. It is natural to become "psyched-up" for a test in contemplation of what is to come. A little emotion can provide a healthy flow of adrenaline to peak your senses and hone your mental ability. This improves your performance on the test and is a very different reaction from fear.

Most technician's fears and insecurities experienced before a test are due to a lack of self-confidence. Those who have not scored well on previous tests or have no confidence in their preparation are those most likely to fail. Be confident that you will do well on your test and your fears should vanish. You will know that you have done everything possible to realize your potential.

Getting Rid of Fear

If you have previously experienced fear of taking a test, it may be difficult to change your attitude immediately. It may be easier to cope with fear if you have a better understanding of what the test is about. A test is merely an assessment of how much the technician knows about a particular task area. Tests, then, are much less threatening when thought of in this manner. This does not mean, however, that you should lower your self-esteem simply because you performed poorly on a test.

You can consider the test essentially as a learning device, providing you with valuable information to evaluate your performance and knowledge. Recognize that no one is perfect. All humans make mistakes. The idea, then, is to make mistakes before the test, learn from them, and avoid repeating them on the test. Fortunately, this is not as difficult as it seems. Practical questions in this study guide include the correct answers to consider if you have made mistakes on the practice test. You should learn where you went wrong so you will not repeat them in the ASE test. If you learn from your mistakes, the stage is set for future growth.

If you understood everything presented up until now, you have the knowledge to become a test-wise technician, but more is required. To be a test-wise technician, you not only have to practice these principles, you have to diligently study in your task area.

Effective Study

The fundamental and vital requirement to induce effective study is a genuine and intense desire to achieve. This is more basic than any rule or technique that will be given here. The key requirement, then, is a driving motivation to learn and to achieve.

If you wish to study effectively, first develop a desire to master your studies and sincerely believe that you will master them. Everything else is secondary to such a desire.

First, build up definite ambitions and ideals toward which your studies can lead. Picture the satisfaction of success. The attitude of the technician may be transformed from merely getting by to an earnest and energetic effort. The best direct stimulus to change may involve nothing more than the deliberate planning of your time. Plan time to study.

Another drive that creates positive study is an interest in the subject studied. As an automotive technician, you can develop an interest in studying particular subjects if you follow these four rules:

1. Acquire information from a variety of sources. The greater your interest in a subject, the easier it is to learn about it. Visit your local library and seek books on the subject you are studying. When you find something new or of interest, make inexpensive photocopies for future study.

2. Merge new information with your previous knowledge. Discover the relationship of new facts to old known facts. Modern developments in automotive technology take on new interest when they are seen in relation to present knowledge.

3. Make new information personal. Relate the new information to matters that are of concern to you. The information you are now reading, for example, has interest to you as you think about how it can help.

4. Use your new knowledge. Raise questions about the points made by the book. Try to anticipate what the next steps and conclusions will be. Discuss this new knowledge, particularly the difficult and questionable points, with your peers.

You will find that when you study with eager interest, you will discover it is no longer work. It is pleasure and you will be fascinated in what you study. Studying can be like reading a novel or seeing a movie that overcomes distractions and requires no effort or willpower. You will discover that the positive relationship between interest and effort works both ways. Even though you perhaps began your studies with little or no interest, simply staying with it helped you to develop an interest in your studies.

Obviously, certain subject matter studies are bound to be of little or no interest, particularly in the beginning. Parts of certain studies may continue to be uninteresting. An honest effort to master those subjects, however, nearly always brings about some level of interest. If you appreciate the necessity and reward of effective studying, you will rarely be disappointed. Here are a few important hints for gaining the determination that is essential to carrying good conclusions into actual practice.

Make Study Definite

Decide what is to be studied and when it is to be studied. If the unit is discouragingly long, break it into two or more parts. Determine exactly what is involved in the first part and learn that. Only then should you proceed to the next part. Stick to a schedule.

The Urge to Learn

Make clear to yourself the relation of your present knowledge to your study materials. Determine the relevance with regard to your long-range goals and ambitions.

Turn your attention away from real or imagined difficulties as well as other things that you would rather be doing. Some major distractions are thoughts of other duties and of disturbing problems. These distractions can usually be put aside, simply shunted off by listing them in a notebook. Most technicians have found that by writing interfering thoughts down, their minds are freed from annoying tensions.

Adopt the most reasonable solution you can find or seek objective help from someone else for personal problems. Personal problems and worry are often causes of ineffective study. Sometimes there are no satisfactory solutions. Some manage to avoid the problems or to meet them without great worry. For those who may wish to find better ways of meeting their personal problems, the following suggestions are offered:

1. Determine as objectively and as definitely as possible where the problem lies. What changes are needed to remove the problem, and which changes, if any, can be made? Sometimes it is wiser to alter your goals than external conditions. If there is no perfect solution, explore the others. Some solutions may be better than others.

2. Seek an understanding confidant who may be able to help analyze and meet your problems. Very often, talking over your problems with someone in whom you have confidence and trust will help you to arrive at a solution.

3. Do not betray yourself by trying to evade the problem or by pretending that it has been solved. If social problem distractions prevent you from studying or doing satisfactory work, it is better to admit this to yourself. You can then decide what can be done about it.

Once you are free of interferences and irritations, it is much easier to stay focused on your studies.

Concentrate

To study effectively, you must concentrate. Your ability to concentrate is governed, to a great extent, by your surroundings as well as your physical condition. When absorbed in study, you must be oblivious to everything else around you. As you learn to concentrate and study, you must also learn to overcome all distractions. There are three kinds of distractions you may face:

1. Distractions in the surrounding area, such as motion, noise, and the glare of lights. The sun shining through a window on your study area, for example, can be very distracting.

    Some technicians find that, for effective study, it is necessary to eliminate visual distractions as well as noises. Others find that they are able to tolerate moderate levels of auditory or visual distraction.

    Make sure your study area is properly lighted and ventilated. The lighting should be adequate but should not shine directly into your eyes or be visible out of the corner of your eye. Also, try to avoid a reflection of the lighting on the pages of your book.

    Whether heated or cooled, the environment should be at a comfortable level. For most, this means a temperature of 78°F–80°F (25.6°C–26.7°C) with a relative humidity of 45 to 50 percent.

2. Distractions arising from your body, such as a headache, fatigue, and hunger. Be in good physical condition. Eat wholesome meals at regular times. Try to eat with your family or friends whenever possible. Meal time should be your recreational period. Do not eat a heavy meal for lunch, and do not resume studies immediately after eating lunch. Just after lunch, try to get some regular exercise, relaxation, and recreation. A little exercise on a regular basis is much more valuable than a lot of exercise only on occasion.

3. Distractions of irrelevant ideas, such as how to repair the garden gate, when you are studying for an automotive-related test.

The problems associated with study are no small matter. These problems of distractions are generally best dealt with by a process of elimination. A few important rules for eliminating distractions follow.

Get Sufficient Sleep

You must get plenty of rest even if it means dropping certain outside activities. Avoid cutting in on your sleep time; you will be rewarded in the long run. If you experience difficulty going to sleep, do something to take your mind off your work and try to relax before going to bed. Some suggestions that may help include a little reading, a warm bath, a short walk, a conversation with a friend, or writing that overdue letter to a distant relative. If sleeplessness is an ongoing problem, consult a physician. Do not try any of the sleep remedies on the market, particularly if you are on medication, without approval of your physician.

If you still have difficulty studying, a final rule may help. Sit down in a favorable place for studying, open your books, and take out your pencil and paper. In a word, go through the motions.

Arrange Your Area

Arrange your chair and work area. To avoid strain and fatigue, whenever possible, shift your position occasionally. Try to be comfortable; however, avoid being too comfortable. It is nearly impossible to study rigorously when settled back in a large easy chair or reclining leisurely on a sofa.

When studying, it is essential to have a plan of action, a time to work, a time to study, and a time for pleasure. If you schedule your day and adhere to the schedule, you will eliminate most of your efforts and worries. A plan that is followed, then, soon becomes the easy and natural routine of the day. Most technicians find it useful to have a definite place and time to study. A particular table and chair should always be used for study and intellectual work. This place will then come to mean study. To be seated in that particular location at a regular scheduled time will automatically lead you to assume a readiness for study.

Don't Daydream

Daydreaming or mind-wandering is an enemy of effective study. Daydreaming is frequently due to an inadequate understanding of words. Use the Glossary or a dictionary to look up the troublesome word. Another frequent cause of daydreaming is a deficient background in the present subject matter. When this is the problem, go back and review the subject matter to obtain the necessary foundation. Just one hour of concentrated study is equivalent to ten hours with frequent lapses of daydreaming. Be on guard against mind-wandering, and pull yourself back into focus on every occasion.

Study Regularly

A system of regularity in study is believed by many scholars to be the secret of success. The daily time schedule must, however, be determined on an individual basis. You must decide how many hours of each day you can devote to your studies. Few technicians really are aware of where their leisure time is spent. An accurate account of how your days are presently being spent is an important first step toward creating an effective daily schedule.

WEEKLY SCHEDULE							
	SUN	MON	TUES	WED	THU	FRI	SAT
6:00							
6:30							
7:00							
7:30							
8:00							
8:30							
9:00							
9:30							
10:00							
10:30							
11:00							
11:30							
NOON							
12:30							
1:00							
1:30							
2:00							
2:30							
3:00							
3:30							
4:00							
4:30							
5:00							
5:30							
6:00							
6:30							
7:00							
7:30							
8:00							
8:30							
9:00							
9:30							
10:00							
10:30							
11:00							
11:30							

The convenient form is for keeping an hourly record of your week's activities. If you fill in the schedule each evening before bedtime, you will soon gain some interesting and useful facts about yourself and your use of your time. If you think over the causes of wasted time, you can determine how you might better spend your time. A practical schedule can be set up by using the following steps:

Mark your fixed commitments, such as work, on your schedule. Be sure to include classes and clubs. Do you have sufficient time left? You can arrive at an estimate of the time you need for studying by counting the hours used during the present week. An often used formula, if you are taking classes, is to multiply the number of hours you spend in class by two. This provides time for class studies. This is then added to your work hours. Do not forget time allocation for travel.

Fill in your schedule for meals and studying. Use as much time as you have available during the normal workday hours. Do not plan, for example, to do all of your studying between 11:00 pm and 1:00 am. Try to select a time for study that you can use every day without interruption. You may have to use two or perhaps three different study periods during the day.

List the things you need to do within a time period. A one-week time frame seems to work well for most technicians. The question you may ask yourself is: "What do I need to do to be able to walk into the test next week, or next month, prepared to pass?"

Break down each task into smaller tasks. The amount of time given to each area must also be settled. In what order will you tackle your schedule? It is best to plan the approximate time for your assignments and the order in which you will do them. In this way, you can avoid the difficulties of not knowing what to do first and of worrying about the other things you should be doing.

List your tasks in the empty spaces on your schedule. Keep some free time unscheduled so you can deal with any unexpected events, such as a dental appointment. You will then have a tentative schedule for the following week. It should be flexible enough to allow some units to be rearranged if necessary. Your schedule should allow time off from your studies. Some use the promise of a planned recreational period as a reward for motivating faithfulness to a schedule. You will more likely lose control of your schedule if it is packed too tightly.

Keep a Record

Keep a record of what you actually do. Use the knowledge you gain by keeping a record of what you are actually doing so you can create or modify a schedule for the following week. Be sure to give yourself credit for movement toward your goals and objectives. If you find that you can not study productively at a particular hour, modify your schedule so as to correct that problem.

Scoring the ASE Test

You can gain a better perspective about tests if you know and understand how they are scored. ASE's tests are scored by American College Testing (ACT), a non-partial, non-biased organization having no vested interest in ASE or in the automotive industry. Each question carries the same weight as any other question. For example, if there are fifty questions, each is worth 2 percent of the total score. The passing grade is 70 percent. That means you must correctly answer thirty-five of the fifty questions to pass the test.

Understand the Test Results

The test results can tell you:
- where your knowledge equals or exceeds that needed for competent performance, or
- where you might need more preparation.

The test results *cannot* tell you:
- how you compare with other technicians, or
- how many questions you answered correctly.

Your ASE test score report will show the number of correct answers you got in each of the content areas. These numbers provide information about your performance in each area of the test. However, because there may be a different number of questions in each area of the test, a high percentage of correct answers in an area with few questions may not offset a low percentage in an area with many questions.

It may be noted that one does not "fail" an ASE test. The technician that does not pass is simply told "More Preparation Needed." Though large differences in percentages may indicate problem areas, it is important to consider how many questions were asked in each area. Since each test evaluates all phases of the work involved in a service specialty, you should be prepared in each area. A low score in one area could keep you from passing an entire test.

Note that a typical test will contain the number of questions indicated above each content area's description. For example:

Engine Repair (Test A1)

Content Area	Questions	Percent of Test
A. General Engine Diagnosis	17	24%
B. Cylinder Head and Valve Train Diagnosis and Repair	18	26%
C. Engine Block Diagnosis and Repair	18	26%
D. Lubrication and Cooling Systems Diagnosis and Repair	9	13%
E. Fuel, Electrical, Ignition, and Exhaust Systems Inspection and Service	8	11%
Total	*70	100%

Note: *The test could contain up to ten additional questions that are included for statistical research purposes only. Your answers to these questions will not affect your score, but since you do not know which ones they are, you should answer all questions in the test. The five-year Recertification Test will cover the same content areas as those listed above. However, the number of questions in each content area of the Recertification Test will be reduced by about one-half.*

"Average"

There is no such thing as average. You cannot determine your overall test score by adding the percentages given for each task area and dividing by the number of areas. It doesn't work that way because there generally are not the same number of questions in each task area. A task area with twenty questions, for example, counts more toward your total score than a task area with ten questions.

So, How Did You Do?

Your test report should give you a good picture of your results and a better understanding of your task areas of strength and weakness.

If you fail to pass the test, you may take it again at any time it is scheduled to be administered. You are the only one who will receive your test score. Test scores will not be given over the telephone by ASE nor will they be released to anyone without your written permission.

3 Are You Sure You're Ready for A1 Test?

Pretest

The purpose of this pretest is to determine the amount of review that you may require prior to taking the ASE automobile test: Engine Repair (Test A1). If you answer all of the pretest questions correctly, complete the sample test in section 5 along with the additional test questions in section 6.

If two or more of your answers to the pretest questions are wrong, study section 4: An Overview of the System before continuing with the sample test and additional test questions.

The pretest answers and explanations are located at the end of the pretest.

1. When installing room-temperature vulcanizing (RTV) sealer:
   A. the components to be sealed should be washed with an oil-base solvent.
   B. a 1/8-inch (3.2-mm) wide bead of RTV should be placed in the center of the sealing surface.
   C. the RTV bead should be placed on one side of any bolt hole.
   D. the RTV bead should be allowed to dry for 10 minutes before component installation.

2. Valve seats are typically ground to an angle of:
   A. 15 degrees or 20 degrees.
   B. 20 degrees or 30 degrees.
   C. 30 degrees or 45 degrees.
   D. 45 degrees or 60 degrees.

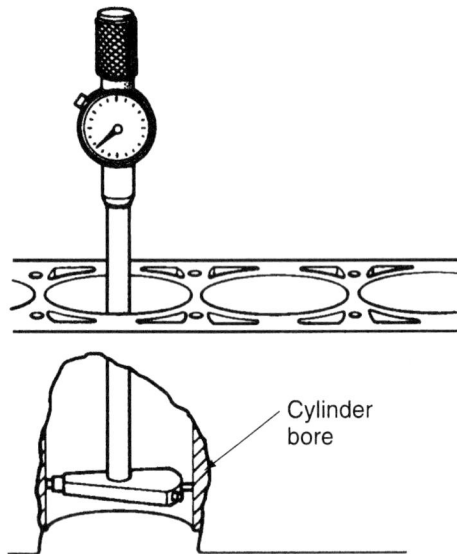

Cylinder bore

3. The gauge in the figure above is used to check:
   A. cylinder wall hardness.
   B. piston clearance.
   C. bore out-of-roundness.
   D. cylinder wall thickness.

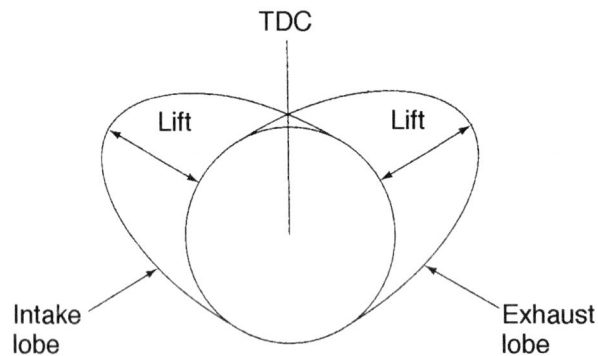

TDC

Lift   Lift

Intake lobe   Exhaust lobe

4. As seen in the figure above, the valve orientation when the piston is at top dead center (TDC) on the exhaust stroke is:
   A. both valves closed.
   B. called valve overlap.
   C. exhaust valve open and intake valve closed.
   D. intake valve open and exhaust valve closed.

5. An intake manifold leak would most likely be noticeable at which of the following?
   A. Low rpm
   B. Cruising speed
   C. Idle
   D. Acceleration

6. When inspecting valve spring retainers, the technician would most likely check for:
   A. roundness.
   B. wear.
   C. scoring.
   D. damage.

7. Technician A says the vibration damper counterbalances the back-and-forth twisting motion of the crankshaft each time a cylinder fires. Technician B says if the seal contact area on the vibration damper hub is scored, the damper assembly must be replaced. Who is right?
   A. A only
   B. B only
   C. Both A and B
   D. Neither A nor B

8. Technician A says the positive crankcase ventilation (PCV) system may draw unfiltered air into the engine through a leaking rocker arm cover gasket. Technician B says the PCV system may draw unfiltered air into the engine through a loose oil filler cap. Who is right?
   A. A only
   B. B only
   C. Both A and B
   D. Neither A nor B

9. An excessively high coolant level in the recovery reservoir may be caused by any of the following problems EXCEPT:
   A. restricted radiator tubes.
   B. a thermostat that is stuck open.
   C. a loose water pump impeller.
   D. an inoperative electric-drive cooling fan.

10. All of the following statements about battery charging are true EXCEPT:
    A. Oxygen gas is produced in the battery cells during the charging process.
    B. Hydrogen gas is produced in the battery cells during the charging process.
    C. The battery is completely charged when the specific gravity reaches 1.225.
    D. Do not charge a maintenance-free battery if the hydrometer in the battery top indicates yellow.

11. During a compression test, a cylinder has 40 percent of the specified compression reading. When the technician performs a wet test, the compression reading on the cylinder is still 40 percent of the specified reading. The LEAST likely cause of the low compression reading is:
    A. a burned exhaust valve.
    B. worn piston rings.
    C. a bent intake valve.
    D. a worn camshaft lobe.

12. Which of the following is the LEAST likely oil pump component measurement?
    A. Outer rotor to housing clearance
    B. Clearance between the rotors
    C. Inner and outer rotor thickness
    D. Inner rotor diameter

Answers to the Test Questions for the Pretest

1. B, 2. C, 3. C, 4. B, 5. C, 6. D, 7. A, 8. C, 9. B, 10. C, 11. B, 12. D

Explanations to the Answers for the Pretest

Question #1
Answer A is wrong. Chlorinated solvent must be used to clean RTV-sealed components.
Answer B is correct. The 1/8-inch (3.2-mm) wide RTV bead should be placed in the center of the sealing surface.
Answer C is wrong. The RTV bead must surround any bolt holes.
Answer D is wrong. RTV sealer starts to cure in five minutes; therefore, components using this sealer must be assembled quickly.

Question #2
Answer A is wrong.
Answer B is wrong.
Answer C is correct. Valve seats are typically ground to an angle of 30 to 45 degrees.
Answer D is wrong.

Question #3
Answer A is wrong. Cylinder wall hardness is not checked with this tool.
Answer B is wrong. Piston clearance is not checked with this tool.
Answer C is correct.
Answer D is wrong. Cylinder wall thickness is not checked with this tool.

Question #4
Answer A is wrong.
Answer B is correct. This is referred to as valve overlap.
Answer C is wrong.
Answer D is wrong.

Question #5
Answer A is wrong. At low rpm the engine is creating moderate vacuum.
Answer B is wrong. At cruising speed the engine is creating moderate vacuum.
Answer C is correct.
Answer D is wrong. During acceleration the engine is creating low vacuum.

Question #6
Answer A is wrong. Roundness is least likely to be checked for by the technician.
Answer B is wrong. Wear is likely to be checked for after damage.
Answer C is wrong. Scoring is likely to be checked for after damage.
Answer D is correct.

Question #7
Answer A is correct. The vibration damper counterbalances the back-and-forth twisting crankshaft motion.
Answer B is wrong. If the damper seal contact area is scored, the damper hub may be machined and a sleeve installed to provide a new seal contact area.
Answer C is wrong.
Answer D is wrong.

Question #8
Answer A is wrong. The PCV system causes a slight vacuum in the engine crankcase. The vacuum will pull unfiltered air into the engine crankcase at a leaking valve cover gasket.
Answer B is wrong. The PCV system causes a slight vacuum in the engine crankcase. The vacuum will pull unfiltered air into the engine crankcase at a loose oil filler cap.
Answer C is correct.
Answer D is wrong.

Question #9
Answer A is wrong. Restricted radiator tubes will decrease the cooling effectiveness of the radiator and cause excessive expansion of the coolant.
Answer B is correct. A stuck open thermostat will prevent the engine coolant from warming up.
Answer C is wrong. If the water pump impeller is loose, it will inhibit coolant flow, prohibiting the cooling effects of the radiator and causing the engine to overheat.
Answer D is wrong. An inoperative electric-drive cooling fan will inhibit air flow past the radiator, prohibiting the cooling effects of the radiator and causing the engine to overheat.

Question #10
Answer A is wrong. Oxygen is produced in the battery cells during the charging process.
Answer B is wrong. Hydrogen is produced in the battery cells during the charging process.
Answer C is correct. A specific gravity reading of 1.225 indicates that the battery is not completely charged.
Answer D is wrong. A yellow indicator in the top of a maintenance-free battery indicates that the battery is low on electrolyte and should not be charged.

Question #11
Answer A is wrong. A burned exhaust valve will cause a low compression reading, which would not be affected by a wet test.
Answer B is correct. Worn piston rings would cause a low compression reading but would most likely be affected by a wet test.
Answer C is wrong. A bent intake valve will cause a low compression reading, which would not be affected by a wet test.
Answer D is wrong. A worn camshaft lobe will cause a low compression reading, which would not be affected by a wet test.

Question #12
Answer A is wrong.
Answer B is wrong.
Answer C is wrong.
Answer D is correct. Inner rotor diameter is least likely component measurement to be made to an oil pump. Oil pump components are most likely to be measured for clearance and thickness.

Types of Questions

ASE certification tests are often thought of as being tricky. They may seem to be tricky if you do not completely understand what is being asked. The following examples will help you recognize certain types of ASE questions and avoid common errors.

Each test is made up of forty to eighty multiple-choice questions. Multiple-choice questions are an efficient way to test knowledge. To answer them correctly, you must think about each choice as a possibility, and then choose the one that best answers the question. To do this, read each word of the question carefully. Do not assume you know what the question is about until you have finished reading it.

Multiple-Choice Questions

One type of multiple-choice question has three wrong answers and one correct answer. The wrong answers, however, may be almost correct, so be careful not to jump at the first answer that seems to be correct. If all the answers seem to be correct, choose the answer that is the most correct. If you readily know the answer, this kind of question does not present a problem. If you are unsure of the answer, analyze the question and the answers. For example:

Question 1:

A rocker panel is a structural member of which vehicle construction type?

A. Front-wheel drive

B. Pickup truck

C. Unibody

D. Full-frame

Analysis:

This question asks for a specific answer. By carefully reading the question, you will find that it asks for a construction type that uses the rocker panel as a structural part of the vehicle.

Answer A is wrong. Front-wheel drive is not a vehicle construction type.

Answer B is wrong. A pickup truck is not a type of vehicle construction.

Answer C is correct. Unibody design creates structural integrity by welding parts together, such as the rocker panels, but does not require exterior cosmetic panels installed for full strength.

Answer D is wrong. Full frame describes a body-over-frame construction type that relies on the frame assembly for structural integrity.

Therefore, the correct answer is C. If the question was read quickly and the words "construction type" were passed over, answer A may have been selected.

EXCEPT Questions

Another type of question used on ASE tests has answers that are all correct except one. The correct answer for this type of question is the answer that is wrong. The word "EXCEPT" will always be in capital letters. You must identify which of the choices is the wrong answer. If you read quickly through the question, you may overlook what the question is asking and answer the question with the first correct statement. This will make your answer wrong. An example of this type of question and the analysis is as follows:

Question 2:
    All of the following are tools for the analysis of structural damage EXCEPT:
    A. height gauge.
    B. tape measure.
    C. dial indicator.
    D. tram gauge.
Analysis:
The question really requires you to identify the tool that is not used for analyzing structural damage. All tools given in the choices are used for analyzing structural damage except one. This question presents two basic problems for the test-taker who reads through the question too quickly. It may be possible to read over the word "EXCEPT" in the question or not think about which type of damage analysis would use answer C. In either case, the correct answer may not be selected. To correctly answer this question, you should know what tools are used for the analysis of structural damage. If you cannot immediately recognize the incorrect tool, you should be able to identify it by analyzing the other choices.

Answer A is wrong. A height gauge *may* be used to analyze structural damage.

Answer B is wrong. A tape measure may be used to analyze structural damage.

Answer C is correct. A dial indicator may be used as a damage analysis tool for moving parts, such as wheels, wheel hubs, and axle shafts, but would not be used to measure structural damage.

Answer D is wrong. A tram gauge *is* used to measure structural damage.

Technician A, Technician B Questions

The type of question that is most popularly associated with an ASE test is the "Technician A says... Technician B says... Who is right?" type. In this type of question, you must identify the correct statement or statements. To answer this type of question correctly, you must carefully read each technician's statement and judge it on its own merit to determine if the statement is true.

Typically, this type of question begins with a statement about some analysis or repair procedure. This is followed by two statements about the cause of the problem, proper inspection, identification, or repair choices. You are asked whether the first statement, the second statement, both statements, or neither statement is correct. Analyzing this type of question is a little easier than the other types because there are only two ideas to consider although there are still four choices for an answer.

Technician A... Technician B questions are really double-true-false questions. The best way to analyze this kind of question is to consider each technician's statement separately. Ask yourself, is A true or false? Is B true or false? Then select your answer from the four choices. An important point to remember is that an ASE Technician A... Technician B question will never have Technician A and B directly disagreeing with each other. That is why you must evaluate each statement independently. An example of this type of question and the analysis of it follows.

Question 3:
    Structural dimensions are being measured. Technician A says comparing measurements from one side to the other is enough to determine the damage. Technician B says a tram gauge can be used when a tape measure cannot measure in a straight line from point to point. Who is right?
    A. A only
    B. B only
    C. Both A and B
    D. Neither A nor B

Analysis:

With some vehicles built asymmetrically, side-to-side measurements are not always equal. The manufacturer's specifications need to be verified with a dimension chart before reaching any conclusions about the structural damage.

Answer A is wrong. Technician A's statement is wrong. A tram gauge would provide a point-to-point measurement when a part, such as a strut tower or air cleaner, interrupts a direct line between the points.

Answer B is correct. Technician B is correct. A tram gauge can be used when a tape measure cannot be used to measure in a straight line from point to point.

Answer C is wrong. Since Technician A is not correct, C cannot be the correct answer.

Answer D is wrong. Since Technician B is correct, D cannot be the correct answer.

Questions with a Figure

About 10 percent of ASE questions will have a figure, as shown in the following example:

Question 4:

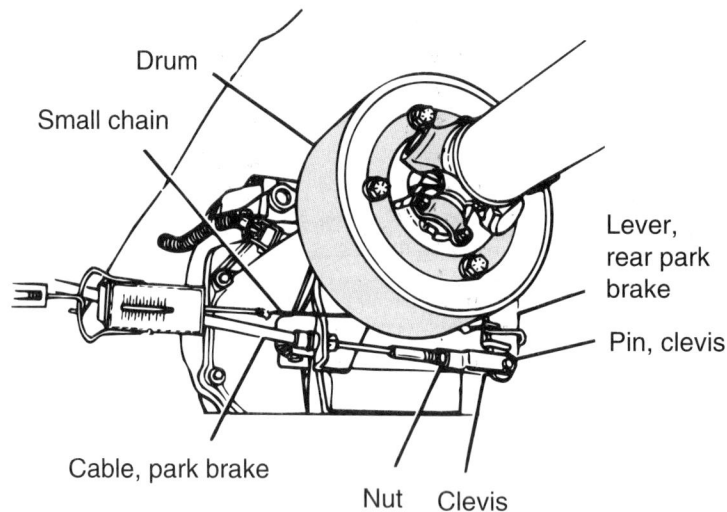

In the measurement shown in the figure above:
A. the driveshaft center support bearing wear is measured.
B. the propeller shaft parking brake is adjusted.
C. the brake shoes must be adjusted before this measurement is taken.
D. the parking brake is released during this measurement.

Analysis:

Answer A is wrong. The driveshaft center support bearing wear is not being measured in this figure.

Answer B is correct. The propeller shaft parking brake adjustment is being performed.

Answer C is wrong. The brake shoes need not be adjusted before adjusting the propeller shaft parking brake.

Answer D is wrong. The parking brake is applied during the propeller shaft parking brake adjustment procedures.

Most-Likely Questions

Most-likely questions are somewhat difficult because only one choice is correct while the other three choices are nearly correct. An example of a most-likely-cause question is as follows:

Question 5:

The most likely cause of reduced turbocharger boost pressure may be a:

A. westgate valve stuck closed.

B. westgate valve stuck open.

C. leaking westgate diaphragm.

D. disconnected westgate linkage.

Analysis:

Answer A is wrong. A westgate valve stuck closed increases turbocharger boost pressure.

Answer B is correct. A westgate valve stuck open decreases turbocharger boost pressure.

Answer C is wrong. A leaking westgate valve diaphragm increases turbocharger boost pressure.

Answer D is wrong. A disconnected westgate valve linkage will increase turbocharger boost pressure.

LEAST-Likely Questions

Notice that in most-likely questions there is no capitalization. This is not so with LEAST-likely type questions. For this type of question, look for the choice that would be the least likely cause of the described situation. Read the entire question carefully before choosing your answer. An example is as follows:

Question 6:

What is the LEAST likely cause of a bent pushrod?

A. Excessive engine speed

B. A sticking valve

C. Excessive valve guide clearance

D. A worn rocker arm stud

Analysis:

Answer A is wrong. Excessive engine speed may cause a bent pushrod.

Answer B is wrong. A sticking valve may cause a bent pushrod.

Answer C is correct. Excessive valve clearance will not generally cause a bent pushrod.

Answer D is wrong. A worn rocker arm stud may cause a bent pushrod.

Summary

There are no four-part multiple-choice ASE questions having "none of the above" or "all of the above" choices. ASE does not use other types of questions, such as fill-in-the-blank, completion, true-false, word-matching, or essay. ASE does not require you to draw diagrams or sketches. If a formula or chart is required to answer a question, it is provided for you. There are no ASE questions that require you to use a pocket calculator.

Testing Time Length

An ASE test session is four hours and fifteen minutes. You may attempt from one to a maximum of four tests in one session. It is recommended, however, that no more than a total of 225 questions be attempted at any test session. This will allow for just over one minute for each question.

Visitors are not permitted at any time. If you wish to leave the test room, for any reason, you must first ask permission. If you finish your test early and wish to leave, you are permitted to do so only during specified dismissal periods.

Monitor Your Progress

You should monitor your progress and set an arbitrary limit to how much time you will need for each question. This should be based on the number of questions you are attempting. It is suggested that you wear a watch because some facilities may not have a clock visible to all areas of the room.

Registration

Test centers are assigned on a first-come, first-served basis. To register for an ASE certification test, you should enroll at least six weeks before the scheduled test date. This should provide sufficient time to assure you a spot in the test center. It should also give you enough time for study in preparation for the test. Test sessions are offered by ASE twice each year, in May and November, at over six hundred sites across the United States. Some tests that relate to emission testing also are given in August in several states.

To register, contact Automotive Service Excellence/American College Testing at:

ASE/ACT
P.O. Box 4007
Iowa City, IA 52243

4 An Overview of the System

Engine Repair (Test A1)

The following section includes the task areas and task lists for this test and a written overview of the topics covered in the test.

The task list describes the actual work you should be able to do as a technician that you will be tested on by the ASE. This is your key to the test and you should review this section carefully. We have based our sample test and additional questions upon these tasks, and the overview section will also support your understanding of the task list. ASE advises that the questions on the test may not equal the number of tasks listed; the task lists tell you what ASE expects you to know how to do and be ready to be tested upon.

At the end of each question in the Sample Test and Additional Test Questions sections, a letter and number will be used as a reference back to this section for additional study. Note the following example: **C.2**.

Task List

C. Engine Block Diagnosis and Repair (18 Questions)

Task 2 Visually inspect engine block for cracks, corrosion, passage condition, core and gallery plug holes, and surface warpage; determine necessary action.

Example:

29. Technician A says a warped cylinder head mounting surface on an engine block may cause valve seat distortion. Technician B says a warped cylinder head mounting surface on an engine block may cause coolant and combustion leaks. Who is right?
    A. A only
    B. B only
    C. Both A and B
    D. Neither A nor B   (C.2)

Question #29
Answer A is wrong.
Answer B is wrong.
Answer C is correct. A warped cylinder head mounting surface on the cylinder block will cause a cylinder head to bend as it is bolted down. This may cause the valve seats to distort. It may also allow coolant and combustion gases to leak past the head gasket.
Answer D is wrong.

Task List and Overview

A. General Engine Diagnosis (17 Questions)

Task 1 Verify driver's complaint and/or road test vehicle; determine necessary action.

The technician must be familiar with basic diagnostic procedures such as the following:

- Listening carefully to the customer's complaint, and questioning the customer in order to obtain more information regarding the problem.
- Identifying the complaint; road testing the vehicle if necessary.
- Thinking of the possible causes of the problem.
- Performing diagnostic tests to locate the exact cause of the problem and always starting with the easiest, quickest test.
- Being sure the customer's complaint is eliminated; road testing the vehicle if necessary.

Task 2 Determine if no-crank, no-start, or hard starting condition is an ignition system, cranking system, fuel system, or engine mechanical problem.

If the starter fails to crank the engine, the problem may range from a faulty starter motor to broken components inside the engine. If no sounds come from the starter motor when it is activated, first disable the ignition system. Then attempt to rotate the crankshaft pulley by hand in the normal direction of rotation. If the crankshaft can be rotated freely through two complete revolutions, the starter is probably defective.

If you are unable to rotate the crankshaft by hand, the engine may be hydrostatically locked or have broken internal components. To check for hydrostatic lock, remove all the spark plugs and attempt to rotate the crankshaft again. If oil or coolant squirts from the spark plug holes, this indicates a bad head gasket, warped cylinder head or block, or a cracked cylinder head or block.

If the crankshaft cannot be rotated at all with the spark plugs removed, or cannot be rotated through at least one complete revolution, the engine may be seized or have broken internal parts. Pull the dipstick and check crankcase oil level. If oil does not register on the dipstick, it is likely that the pistons are seized in their bores or the connecting rods are seized to the crankshaft. If the oil level is sufficient, a broken component may have lodged between moving parts inside the cylinder block, preventing the parts from rotating.

Many overhead camshaft (OHC) engines are non-freewheeling or "interference" engines. On these engines, a no-crank condition may be caused by piston-to-valve contact. This is a common occurrence when a timing belt slips or breaks, but it may also occur on engines fitted with a timing chain and sprockets. On many belt-driven OHC engines it is possible to easily loosen or remove part of the timing belt cover. Do this, if possible, and check for obvious signs of belt failure.

If the customer states that the starter cranks the engine but it will not start (or takes a long time to start), confirm that the valve train is operating properly before attempting to crank the engine yourself. If the timing belt or chain is broken or jumped, additional cranking may cause severe engine damage.

A no-start or hard starting complaint can be caused by a faulty ignition, fuel, or emission control system. These complaints can also be caused by broken or slipped valve train timing components, especially on free-wheeling engines. A broken timing device may cause some cylinders to have good compression while others have none. A slipped timing device may result in all cylinders having low compression. To determine if the belt or chain is functioning properly, rotate the crankshaft by hand while observing the distributor rotor or camshaft. If these components fail to rotate with the crankshaft, the timing belt or chain is broken. If they do rotate with the crankshaft, confirm proper indexing of the rotor or camshaft to determine if the belt or chain has slipped. Rotate the crankshaft until the piston in cylinder #1 is at TDC on the compression stroke. Then check distributor rotor or camshaft position to make sure that it is correct.

If the engine mechanical system is okay, check the ignition system. First, reconnect or install any components used to disable the system previously. Then disconnect one of the spark plug wires and install a spark testing tool. Crank the engine and verify that a strong spark occurs at the tool. If no sparks occur, begin ignition system diagnosis.

Check the fuel system next. On carbureted engines, remove the air cleaner and inspect the carburetor venturi area while operating the throttle linkage. When the throttle is opened quickly, a stream(s) of liquid gasoline should spray from the accelerator pump nozzle(s) into the venturi area. If this does not occur, the accelerator pump circuit is faulty or the carburetor bowl is not receiving fuel. Proceed with fuel system diagnosis.

Fuel injected engines usually receive high-pressure fuel from an electric pump. To verify that the fuel pump is operating and fuel is reaching the engine, locate the fuel line that supplies fuel to the throttle body or fuel injector rail. Carefully disconnect the fuel line (place a shop towel under the line to absorb any fuel that leaks out) and install an in-line fuel pressure gauge. Crank the engine and verify that the gauge registers adequate pressure (about 40 psi). If the gauge does not register any pressure or registers very low pressure, proceed with fuel system diagnosis. If fuel pressure is adequate, begin diagnosis of the fuel injection control system.

Task 3   Inspect engine assembly for fuel, oil, coolant, and other leaks; determine necessary action.

The source of fluid leaks can be difficult to locate. Determining what type of fluid is leaking will reduce the number of possible leak locations.

Engine oil usually leaks from faulty gaskets and seals, but it can also leak from cracked castings, faulty pressure switches or sending units, and loose tapered oil gallery plugs. Oil can leak from an area high on the engine (like a V-type engine intake manifold rear seal) and run down the engine, appearing at the rear of the oil pan. Do not assume that the "wet" area is the source of the leak. Clean the area and run the engine to check for fresh fluid. Also, do not immediately assume that a leaking seal or gasket is faulty. Excessive blowby or a faulty PCV system can pressurize the crankcase, forcing oil past a seal or gasket that is in good condition. Hard-to-find oil leaks can be located by pouring a small quantity of fluorescent dye into the crankcase and running the engine. When an ultraviolet light is shined onto the engine, oil containing the dye will glow to reveal the leak point.

Fuel may leak from loose connections or damaged components. Check for loose hose clamps and fuel line fitting nuts. Check hoses for swelling, cracks, and damage from abrasion. Check metal lines for cracks and corrosion. Fuel injected engines usually receive high-pressure fuel from an in-tank electric pump and have additional components. Check flexible hoses with crimped-on metal fittings for swelling or bulges. Check for leaking O-ring connections and a leaking fuel pressure regulator (often mounted on the throttle body unit or injector rail).

Corroded core or "freeze out" plugs are common coolant leakage points, as are faulty hoses and water pumps. Check coolant temperature sensors, sending units, and thermal vacuum switches, too. Some engines have core plugs at the back of the cylinder block and/or head. If the engine consumes coolant, but you cannot find evidence of leaking coolant, check the engine oil level and condition. Coolant may be leaking into the crankcase, causing sludge to form. Coolant can also leak into the combustion chambers after the engine is shut down.

On vehicles equipped with power steering, check the fluid level. Leaking power steering fluid may be mistaken for engine oil or transmission fluid.

Task 4   Listen to engine noises; determine necessary action.

Different types of engine part failures often make distinctive sounds. First, be sure that the noise is actually coming from the engine. A faulty water pump, alternator, power steering pump, A/C compressor or air injection pump can make noises that appear to be coming from inside the engine. Loose or broken accessory mounting brackets can also cause noises that sound like engine internal problems. Listen to each of the accessories using a stethoscope to determine if it is the source of a noise. If in doubt, temporarily remove the drive belt from an accessory to prevent it from operating.

A faulty crankshaft main or rod bearing usually makes a knocking sound that is very deep in pitch. Main bearing knock is usually a thumping noise most noticeable when the engine is first started. Connecting rod bearings also cause a heavy knocking sound, and engine oil pressure may also be low, especially at idle. When the faulty cylinder is disabled during a cylinder balance test, the knocking sound will diminish. Loose flywheel bolts may cause a thumping noise at idle. Camshaft bearings usually do not cause a noise unless severely worn.

Worn pistons and cylinders cause a rapping noise while accelerating. When performing a cylinder balance test, piston noise can increase when the faulty cylinder is disabled (the opposite reaction of a bad connecting rod bearing). A piston pin with excessive clearance often makes a "double click" noise when the engine is idling.

Lifters also make a distinctive noise, a loud ticking sound. One way to isolate lifter (or other valve train) noise from connecting rod noise is to remember that the camshaft operates at half of crankshaft speed. It is common for a lifter with excessive leak-down to tick for a few seconds after the engine starts. The noise goes away once full oil pressure is developed.

Task 5   Diagnose the cause of excessive oil consumption, coolant consumption, unusual engine exhaust color, odor, and sound; determine necessary action.

Excessive oil consumption can be due to oil leaking from the engine or oil being drawn into the cylinders and burned. Before blaming internal components, be absolutely sure that oil is not leaking from the engine. In some cases oil leaks only when the engine is running. If necessary, raise the vehicle on a lift while the engine is running to check for leaks.

Oil can enter the cylinders several different ways, including: worn rings; scored cylinder walls; worn valve guides, seals, and stems; worn turbocharger seals; and plugged oil drain passages. As a general rule, an engine that is "burning oil" will emit blue-gray exhaust. This may be more noticeable on acceleration and deceleration. Do not confuse the blue-gray smoke due to oil consumption with the black exhaust that occurs when the air/fuel ratio is too rich.

Plugged oil drain passages in the cylinder head or block can cause excessive oil consumption even when rings and guides are in good condition. To check for this, remove the oil filler cap or another component fitted to a valve cover and start the engine. If the oil level inside the cover rises steadily as the engine runs and reaches the top of the valve guides, the drain passages are clogged. While these passages can usually be cleared of sludge, the sludge is an indication that the engine was poorly maintained. Clearing the passages will probably reduce oil consumption, but the engine may experience other problems in the near future.

Perform compression tests (Task A.8) and cylinder leakage tests (Task A.9) to confirm that piston rings/cylinders or valve guides are worn.

If the engine is turbocharged, first perform oil consumption diagnosis as though the engine was *not* turbocharged. While turbos are commonly blamed for excessive oil consumption problems, about half of the turbos returned under warranty are not defective. If oil is found in the turbo compressor housing or intake manifold, check the oil drain from the turbo housing to the block. If it is obstructed, oil under pressure will be forced into the engine. Check the PCV system, too. If the PCV valve does not close during "boost" conditions, the crankcase will be pressurized. This may pressurize the turbo oil drain passage, forcing oil into the turbo housing.

Like oil consumption, coolant consumption may be caused by coolant leaking from the cooling system or coolant leaking into the engine (or passenger compartment). First, eliminate external leaks as the cause of coolant consumption by performing cooling system pressure tests (Task D.3). If cooling system pressure drops during the tests, but no leaks are found, check the engine oil level and condition. Leaks into the crankcase will raise the oil level and cause severe sludging. If coolant is being drawn into the combustion chambers, the exhaust will be gray or white. The engine will continue to emit this smoke long after the time it usually takes for moisture to be purged from the exhaust system. On vehicles equipped with electronically controlled carburetors or fuel injection, coolant passing through the exhaust system will "poison" the oxygen sensor.

Some engine problems can be diagnosed by listening to the exhaust pulses at the tailpipe. If all cylinders are firing properly, the exhaust should consist of steady pulses. A puffing noise that occurs at regular intervals usually indicates a cylinder misfire caused by a compression, ignition, or fuel system defect. Puffing noises that occur erratically are usually caused by ignition or fuel system defects. Engine idle speed may also be unsteady.

A high-pitched squealing noise during hard acceleration may be caused by a small leak in the exhaust system, particularly in the exhaust manifolds or exhaust pipe. The leak may also be noticeable at idle as a ticking noise.

Another common engine noise is a high-pitched whistle at idle and low engine speeds. Check for vacuum leaks at the intake manifold gaskets. Also check for cracked or disconnected vacuum hoses. A vacuum leak whistle gradually decreases when the engine is accelerated and the intake vacuum decreases.

A strong sulfur, or rotten egg, smell coming from the exhaust system of a car fitted with a catalytic converter may indicate a rich air/fuel ratio.

Task 6   Perform engine vacuum tests; determine necessary action.

A vacuum test can be used to help pinpoint the cause of an engine problem. The vacuum gauge should be connected directly to the intake manifold.

On an engine that is performing correctly, the vacuum gauge reading should be between 17 and 22 in. Hg (45 and 28 kPa absolute) and steady with the engine idling. Some abnormal vacuum gauge readings and typical problems associated with them are listed below.

- A slightly low but steady reading indicates late ignition timing.
- A very low but steady reading indicates that the intake manifold has a significant leak.
- When the vacuum gauge pointer drifts back and forth between approximately 11 and 16 in. Hg (67 and 48 kPa absolute) on an idling carbureted engine, the idle mixture screws should be adjusted. On a fuel-injected engine, the injectors require cleaning or replacing.
- Burned or leaking valves cause the vacuum gauge to fluctuate between 12 and 18 in. Hg (62 and 41 kPa absolute).
- Weak valve springs result in a vacuum gauge fluctuation between 10 and 25 in. Hg (69 and 17 kPa absolute).
- A leaking head gasket may cause a vacuum gauge fluctuation between 7 and 20 in. Hg (79 and 35 kPa absolute).
- If the valves are sticking, the vacuum gauge fluctuates between 14 and 18 in. Hg (55 and 41 kPa absolute).
- If, when the engine is accelerated and held steady at a higher speed, the vacuum gauge pointer gradually rises, the catalytic converter or other exhaust system components are restricted.

Task 7   Perform cylinder power balance tests; determine necessary action.

If all the cylinders in an engine are functioning properly, each one contributes the same amount of power. Therefore, disabling, or "knocking out," the ignition spark in the cylinders, one at a time, should cause engine speed to drop the same amount for each cylinder. If engine speed drops very little when a particular cylinder is disabled, that cylinder is not contributing the same amount of power as the other cylinders. When this is the case, the technician must determine which of the engine systems is at fault: mechanical (compression), ignition, or fuel (including the air intake system).

Mechanical systems can be checked by performing compression and cylinder leakage tests (Tasks A.8 and A.9). Ignition system components can be checked visually and, if no problem is obvious, with an engine analyzer or oscilloscope. The fuel system on carbureted and throttle body injected engines does not usually affect a single cylinder. Instead, check for vacuum leaks. An intake manifold vacuum leak usually causes the affected cylinder to misfire at low engine speeds, especially at idle. At higher speeds and engine loads, the misfire may disappear. On port fuel injected engines, plugged, restricted, and leaking injectors are a common problem. Test equipment is available which allows the technician to momentarily energize each injector and measure the resulting fuel system pressure drop. An injector that causes a pressure drop that is smaller or larger than that of the other injectors should be cleaned and retested. If the pressure drop is still smaller or larger, the injector should be replaced.

Task 8   Perform cylinder compression tests; determine necessary action.

The ignition and fuel injection system must be disabled before proceeding with the compression test. During the compression test, the throttle is blocked open and the engine is cranked through four compression strokes for each cylinder. The compression readings are recorded for each stroke and compared to the manufacturer's specifications. Slightly low compression readings in all cylinders are not cause for concern if engine performance is acceptable. Compression readings that vary more than 20 percent (from the highest to the lowest) are cause for concern. Compression readings may be interpreted as follows:
- When the compression readings on all the cylinders are about equal, but significantly lower than specifications, the piston rings or cylinder walls are probably worn. If compression in all cylinders is low and the engine spins freely during cranking, check the valve timing. The timing belt or sprocket may have jumped.
- Low compression readings on one or more cylinders indicates worn rings, leaking valves, a blown head gasket, or a cracked cylinder head. Performing a "wet" test will narrow down the cause of the problem. Remove the compression tester and squirt approximately two or three teaspoons of engine oil through the spark plug opening into the cylinder having the low compression reading. Crank the engine to distribute the oil around the cylinder wall and then retest the compression. If the compression reading improves considerably, the rings (or cylinders) are worn. If compression does not increase, the valves are leaking, the head gasket is blown, or the cylinder head is cracked.
- Low compression readings in two adjacent cylinders is probably due to a leaking head gasket or cracked cylinder head.
- Zero compression in a cylinder is usually caused by a hole in a piston or a severely burned exhaust valve. If the zero compression reading is caused by a hole in the piston, the engine will have excessive blowby.
- Higher than specified compression usually indicates carbon deposits in the combustion chamber.

Task 9   Perform cylinder leakage tests; determine necessary action.

During a cylinder leakage test, a regulated amount of air from the shop air supply is forced into the cylinder while both the exhaust and intake valves are closed. The gauge on the leakage tester indicates the percentage of leakage in the cylinder. A gauge reading of 0 percent indicates that there is no cylinder leakage. If the reading is 100 percent, the cylinder is not holding any air.

If cylinder leakage exceeds 20 percent, check for air escaping from the tailpipe, the positive crankcase ventilation (PCV) valve opening in the rocker arm cover, and the top of the throttle body or carburetor. Air escaping from the tailpipe indicates an exhaust valve leak. When the air is coming out of the PCV valve opening, the piston rings are leaking. An intake valve is leaking if air is escaping from the top of the throttle body or carburetor. Remove the radiator cap and check the coolant for bubbles, which indicates a leaking head gasket or cracked head.

B. Cylinder Head and Valve Train Diagnosis and Repair (18 Questions)

Task 1   **Remove cylinder heads, disassemble, clean, and prepare for inspection according to manufacturer's procedures.**

Remove a cylinder head only when the engine is cold. Removing a warm cylinder head may allow the head to warp, especially if it is made of aluminum.

It is not necessary to remove the cylinder head to service many valve train components. This includes the valve springs, oil seals, retainers, and valve locks. A claw type valve spring compressor can be used to compress the valve spring while the head is mounted on the block.

Cylinder head removal on an overhead valve (OHV) "pushrod" engine requires that the rocker arm cover, rocker arms, and pushrods be removed. When removing a stamped sheet metal rocker cover, be careful to avoid damaging the cover by prying against one of its edges. Covers sealed to the cylinder head by RTV sealant can be loosened by striking a strong area at the end of the cover with a rubber mallet. The object is to get the cover to "slide" slightly on the head to break the RTV seal.

On engines with stud-mounted rocker arms, loosen the lash adjusting nuts enough to turn the rockers sideways and remove the pushrods. On engines with tube-mounted rocker arms, loosen each tube-mounting bolt one turn at a time until the pushrods are loose. Do not fully loosening the bolts one at a time, as this may cause the tube to warp. Remove the tube and rocker assembly, and then the pushrods. Keep all removed valve train parts in some sort of organizer so each can be returned to its original position.

Loosen the intake and exhaust manifold bolts, noting the positions of long bolts, short bolts, bolts with special heads for accessory mounting brackets, etc. Making a diagram showing the locations of these bolts now will simplify reassembly. Remove the intake and exhaust manifolds.

Remove the cylinder head bolts, loosening the bolts in a sequence *opposite* that of the tightening sequence. Again, note and record the positions of special bolts. Remove the cylinder head from the engine. Cylinder heads can be quite heavy, so ask an assistant to help you, especially if the engine is still mounted in the vehicle.

Use a spring compressor to compress the valve springs and then remove the valve locks, or "keepers." Release the compressor and remove the retainer, rotator, spring, and spring seats from the head. Keep all parts in an organizer so they can be returned to their original cylinder. Check the valve stem tips for mushrooming. If it is present, the tip must be dressed with a file before the valves are removed from the head. Remove the valves from the cylinder head and place them in an organizer.

When removing the cylinder head from an overhead camshaft (OHC) design engine, the timing belt or chain must first be disconnected from the camshaft. The procedure for doing this varies from manufacturer to manufacturer. On some engines with a chain-driven camshaft, the camshaft sprocket is unbolted from cam. The cylinder head assembly is then removed, leaving the chain and sprockets in position on the engine. On some engines with a belt-driven camshaft, the belt tensioner is loosened and the belt is slipped off the camshaft sprocket. On other engines the timing cover and belt must be completely removed from the engine. If the timing belt is being removed from the engine and will be reused, mark the direction of rotation on the belt. Reinstall the belt so it rotates in the same direction. Never crank the engine after a timing device has been loosened or removed (until the cylinder head has been removed). Cranking an engine while the timing belt is loose or disconnected can cause immediate and serious engine damage.

The basic cylinder head removal procedure varies from manufacturer to manufacturer. On some OHC engines the cylinder head, camshaft(s), and rocker arms (if used) are removed as an assembly after loosening and removing the cylinder head bolts. On some engines the rocker shaft and arm assembly (including the upper half of the cam bearings) must first be removed to access cylinder head mounting bolts. Refer to the appropriate service manual for information.

With the cylinder head removed from the engine, compress the valve springs and remove the locks, retainers, springs, seals, and valves.

Task 2   Visually inspect cylinder heads for cracks and gasket surface areas for warpage, corrosion, and leakage; check passage condition.

While the cylinder heads from any engine should be carefully inspected, the heads from engines with serious mechanical problems (blown head gasket, coolant consumption, overheating, oil sludging, etc.) should receive special attention. First look at the old head gaskets to determine if a problem area is visible. If one (or more) is, match the area to the contact area on the cylinder head.

Check the cylinder head for cracks, paying special attention to the combustion chambers and the areas between intake and exhaust valves. An electromagnetic-type tester and iron filings (MAGNA-FLUX®) may be used to check for cracks in cast-iron heads. A dye penetrant may be used to locate cracks in aluminum heads. Machine shops can usually locate hard-to-find cracks by pressure testing. In this type of test, all coolant passages are blocked using metal plates. The coolant jacket is then filled with compressed air and the head is submerged in a tank of water. Bubbles escaping from the head reveal leak areas. Cracked cast iron heads are usually replaced since reliable crack repair is time consuming, expensive, and requires special equipment and skills. Cracked aluminum heads are sometimes welded, but this is also a specialty area.

Use a straightedge and a feeler gauge to check the cylinder head for warpage at several locations. Place the straightedge diagonally across the combustion chamber side of the head (both possible directions), along the length of the head (at the middle and each side), and across the head (at each end). In general, a six cylinder head should not be warped more than 0.006 inch (0.152 mm), a four cylinder head should not be warped more than 0.004 inch (0.102 mm), and a three cylinder head should not be warped more than 0.003 inch (0.076 mm). A head should not be warped more than 0.003 inch (0.076 mm) in any 6-inch (152-mm) length, either. Check the manufacturer's service manual for exact specifications. A cylinder head that is excessively warped must be resurfaced.

On overhead camshaft (OHC) engines, be sure to check warpage on the cam side of the head before resurfacing and reinstalling the head. If warpage exceeds 0.002 inch (0.051 mm) over the length of the head, the camshaft will bind, flex, and may break. An aluminum OHC head can often be straightened, but this is a specialty area that must be left to an expert. An excessively warped OHC head should be replaced by a new or remanufactured head.

Inspect the coolant passages in the cylinder head as thoroughly as possible. Shine a flashlight into the passages, looking for corrosion, rust, and trapped debris. A cylinder head that shows evidence of severe pitting in the cooling jacket should be replaced.

Task 3   **Inspect and test valve springs for squareness, pressure, and free height comparison; replace as necessary.**

Valve springs must be measured for free length and squareness by placing each spring against a steel square that is resting on a surface plate. Free length is the height of the valve spring measured with no tension on the spring. Squareness is the vertical straightness of the valve spring. The valve spring must be replaced if it does not have the specified free length. Spring squareness is also checked using a steel square and a surface plate. In one method, the spring is rotated while checking for height variance. A variance of more than 1/16 inch (1.6 mm) indicates that the spring is bent and must be replaced. Alternatively, the spring can be rotated until the top coil is the greatest distance from the square. Compare the measured distance to the specification supplied by the engine manufacturer.

A valve spring tester is used to measure valve spring pressure or tension. Several different types of testers are available, but on most types the tool table is first adjusted to the spring test length specified by the engine manufacturer. The spring is then placed over the stud on the tool table and compressed. Some testers display spring tension directly on a gauge. On others a beam-type torque wrench is used to operate the tester and compress the spring until a click or ping is heard. The torque wrench reading at the instant the click or ping occurs is then multiplied by two to obtain the spring load at test length.

Task 4   **Inspect valve spring retainers, rotators, locks, and valve lock grooves.**

Valve spring retainers and locks must be checked for wear, scoring, or damage. When any of these conditions are present, replace the components.

Valve rotators are best inspected before cylinder head removal and disassembly since they usually cannot be taken apart. Rotators can be located on top of the valve spring (built into the spring retainer) or between the valve spring and the cylinder head. To test a rotator, mark the top of the spring retainer with a dab of paint. Then start the engine and run it at about 1,500 rpm. The retainer should slowly rotate. The direction of rotation is not important. If the retainer does not rotate, replace the rotator.

In some cases, a defective rotator can be diagnosed with the cylinder head removed. Check the valve stem tip wear pattern. A shallow groove or channel running across the tip indicates that the valve has not been rotating. Replace the rotator on any valve with this condition.

Inspect the valve lock grooves machined into the valve stems. Look for damage and wear, particularly for round shoulders. If the shoulders are uneven or rounded, replace the valve. Valve lock failures cause severe engine damage.

Task 5   **Replace valve stem seals.**

Lubricate the valve stems and guides with the manufacturer's recommended engine oil. Then slip each valve into its guide. If the engine has valve spring seats, slip them over the valve guides.

On engines equipped with umbrella type valve seals, slip the new seals over the valve stems. Work carefully to avoid damaging new seals on valve stem lock grooves. A damaged seal will cause excessive oil consumption. Some seals come with an installation tool. The tool is simply a short plastic sleeve that is slipped over the tip of each valve stem before the seal is installed. The sleeve extends down far enough to cover the lock grooves. After installing each seal, push it down against the top of the valve guide and remove the installation tool.

On engines equipped with positive type valve seals, the installation procedure is the same with one important difference. Positive type seals must be pushed down over the top of the valve guide. Each seal has some sort of retaining device to keep it attached to the guide. Some seals have a flat, circular spring that wraps around the seal. Some use garter springs to hold the seal in place. Others have a molded-in ridge on their inner diameter which mates with a groove machined onto the valve guide. Whatever the retaining method, be sure that the valve guide seal is securely attached to the guide. Some manufacturers specify that a special tool should be used to drive the positive seal onto the valve guide. For further information, refer to a service manual for the vehicle.

Task 6   Inspect valve guides for wear; check valve guide height and stem-to-guide clearance; determine needed repairs.

Valve guides should be measured near the top, center, and bottom using a hole gauge. Measure the valve stem diameter with a micrometer in the same three positions, and subtract the stem readings from the guide measurements to obtain the clearance. An alternate method for measuring stem-to-guide clearance is to install the valve in the guide with the valve about 1/8 inch (3.18 mm) off its seat. Mount a dial indicator against the valve margin or against the valve stem below the lock grooves. Move the valve from side to side while observing the clearance reading on the dial indicator. Divide the reading by two to obtain stem-to-guide clearance.

If clearance exceeds specifications, the valve guides may be replaced, knurled, or bored out, and a thin-wall liner installed. Excessive valve stem-to-guide clearance may result in an improper valve seating and lower compression. Increased oil consumption may result from excessive valve stem-to-guide clearance.

Valve guide height is usually measured from the top of the spring seat to the top of the guide. If guide height is not within specifications, check to see if a pressed-in guide was improperly installed or has moved in the cylinder head.

Task 7   Inspect valves; resurface or replace according to manufacturers' procedures.

Inspect the valve faces for cracks, burning, or pitting. Cracked or burned valves must be replaced. Valves with minor pitting can be resurfaced.

Check for a bent valve by rolling the valve stem on a surface plate or chucking the valve stem in a valve grinder. Watch the valve head as the valve spins. If the head wobbles, the valve is bent and must be replaced.

Measure the overall length of each valve. Any valve that is not within the manufacturer's specifications should be replaced. Inspect the surface of the valve tip for wear and score marks. When these conditions are present, resurface the tip on a valve stem grinder or replace the valve. If valve length is at the low end of specifications, grinding the valve stem tip may make the valve too short. Take this into consideration before grinding the tip.

Measure valve stem diameter at the unworn area just below the lock grooves. Then measure stem diameter at the worn area that contacts the top of the valve guide. If the difference is 0.001 inch (0.025 mm) or more, the valve is excessively worn and should be replaced.

Measure the valve margin thickness. When this thickness is less than specified (typically 1/32 inch [0.79 mm]), replace the valve. If resurfacing the valve face will cause the valve margin to measure less than 1/32 inch (0.79 mm), the valve should also be replaced. Installing a valve that has too small a margin (a knife edge) will lead to valve overheating and burning.

Resurface the valve face on a valve grinder. If the specified valve seat angle is 45°, many vehicle manufacturers recommend grinding the valve face to an interference angle of 44.5°. Interference angles promote superior initial seating and act to provide a self-cleaning action particularly important on leaded fuel engines.

Task 8   Inspect and resurface valve seats according to manufacturers' procedures.

Inspect the valve seats for pitting, cracks, and wear. Integral valve seats (those cast as part of the cylinder head) can be resurfaced, if they are not excessively worn. Insert valve seats (those pressed into a cylinder head) are usually found on aluminum cylinder heads. If they are not excessively worn they can also be resurfaced. If severely worn or cracked, they can often be replaced (if the area under the seat is not damaged).

To resurface a valve seat, the proper sized pilot is first placed in the valve guide. A grinding wheel is then placed over this pilot, and the wheel is rotated with an electric drive tool. A grinding stone of the proper size and angle must be installed on the grinding wheel. This grinding wheel must fit on the valve seat without touching any other part of the head surface.

Valve seat angles vary, but most have a sealing surface cut to 45°. On heads having a "three angle valve job," an additional topping angle of 30° is ground above the valve head (toward the combustion chamber) and a throating angle of 60° is cut below the valve head (toward the valve guide). The valve sealing area can be moved up or down in the head by controlling the depth of the 30° and 60° cuts.

An insert-type valve seat may be removed from the head using a special puller or a pry bar. To install a new seat, it should first be chilled (often using dry ice). A special driver is then used to force the new seat into the cylinder head. The insert should be staked in place after installation.

Task 9   Check valve face-to-seat contact and valve seat concentricity (runout).

After grinding the valves and resurfacing the valve seats, check valve face-to-seat contact. Apply blue dye (Prussian blue) to the valve seat and install the valve against the seat. Tap the valve head, but *do not* rotate the valve. Remove the valve and observe the dye on the valve face to determine the width and height of the transfer area (contact area). The width and height of the contact area can be changed by revising the topping or throating angles in the valve seat as noted in Task A.8.

The transfer area also shows whether or not the valve and seat are concentric. If the blue dye transfers 360° around the valve face, the valve and seat are concentric. If the blue dye does not appear 360° around the valve face, replace the valve.

A valve seat concentricity tester containing a dial indicator may also be used to measure valve seat concentricity.

Task 10   Check valve spring installed (assembled) height and valve stem height; service valve and spring assemblies as necessary.

Measure the installed valve stem height from the spring seat surface on the cylinder head to the valve stem tip. If stem installed height is greater than specifications, the valve stem is stretched or too much material has been removed from the valve face or seat. Install a new valve and measure stem height again. If the measurement is still excessive, replace the seat or cylinder head. Excessive valve stem height moves the plunger downward in a hydraulic valve lifter and may cause valve train components to bottom out.

Measure the installed valve spring height from the lower edge of the top retainer to the spring seat. If this measurement is excessive, install shims between the bottom of the valve spring and the top of the spring seat surface on the cylinder head. Excessive installed valve spring height reduces valve spring tension, which may result in valve float and cylinder misfiring at higher speeds.

Task 11   Inspect pushrods, rocker arms, rocker arm pivots, and shafts for wear, bending, cracks, looseness, and blocked oil passages; repair or replace as required.

Pushrods should be inspected for a bent condition and wear on the ends. Roll the pushrod on a level surface to check for a bent condition. Bent pushrods usually indicate interference in the valve train, such as a sticking valve, improper valve adjustment, or mechanical interference due to improper valve timing. If the pushrod has an oil passage to provide oil to the rocker arm, make sure that the passage is not obstructed.

Worn rocker arms, shafts, or pivots cause improper valve adjustment and a clicking noise in the valve train. Check rocker arm shafts for wear and scoring in the rocker arm contact area. Check the shafts for cracks, bending, and loose/leaking oil passage plugs (if fitted). Check rocker arms for scoring at the pivot area and valve stem tip contact area. Worn rocker arms should be replaced.

Task 12   Inspect and replace hydraulic or mechanical lifters/lash adjusters.

When the valve train is serviced, the valve lifters should be removed, cleaned, and inspected. Hydraulic lifters should also be tested for leak-down.

Check the outsides of the lifters for any signs of wear or scoring. Worn or damaged lifters must be replaced, and the cylinder block should also be checked for damage. Check the bottom of conventional (nonroller) lifters for condition and profile. The bottom should be smooth and slightly convex. Replace any lifter with a scored or pitted bottom. Replace any lifter with a flat or concave bottom. If the bottom of the lifter is scored or concave, replace the camshaft, too. Remember, any time a new camshaft is installed, new lifters must also be installed. Check the roller on roller lifters. The roller should be smooth and rotate smoothly and freely without looseness.

Sticking lifter plungers cause a clicking noise, especially when the engine is started. Burned valves may be caused by sticking lifter plungers. If the lifters are cleaned and reassembled, they should be tested in a leak-down tester. This tester checks the time required to bottom the plunger with a specific weight applied to the plunger. When the lifters leak down too quickly, a clicking noise may be heard in the valve train with the engine idling.

Task 13   Adjust valves on engines with mechanical or hydraulic lifters.

Valve lash adjusting procedures and mechanisms vary from manufacturer to manufacturer. On some engines valves are adjusted while the engine is cold. On others, the engine should be at operating temperature. Refer to the appropriate service manual for instructions.

On some engines with mechanical valve lifters, the rocker arms have an adjustment screw and a locknut on the valve stem end of the arm. Other engines use an interference fit screw without a locknut. The adjustment procedure usually involves rotating the crankshaft to position the piston in the cylinder being adjusted at top dead center (TDC) on the compression stroke. Feeler gauges are then inserted between the adjusting screw and the valve stem. If clearance is excessive, the adjustment screw locknut (if equipped) is loosened and the adjustment screw is turned. When clearance is correct, a feeler gauge of the correct thickness will slide between the adjusting screw and the valve stem with a light push fit. Tighten the locknut (if equipped) when clearance is correct.

Some OHC engines fitted with mechanical valve lifters have removable metal pads in each lifter or spring retainer. Clearance is measured by placing feeler gauges between the cam lobe and the lifter or spring retainer while the piston is at TDC on the compression stroke. Pads are available in various thicknesses to adjust the clearance to specifications.

Some valve trains have hydraulic valve lifters and individual rocker arm pivots retained with self-locking nuts. These valve trains require an initial adjustment of the rocker arm nut to position the lifter plunger. With the valve closed, loosen the rocker arm nut until there is clearance between the end of the rocker arm and the valve stem. Slowly turn the rocker arm nut clockwise while rotating the pushrod. Continue rotating the rocker arm nut until the end of the rocker arm contacts the end of the valve stem, and the push rod becomes harder to turn. Continue turning the rocker arm nut clockwise the number of turns specified in the service manual. In some engines, this specification is one turn plus or minus one quarter turn.

The valve train on engines with hydraulic lifters and stud-mounted rocker arms can also be adjusted while the engine is running. After removing the valve covers, install oil shrouds on the rocker arms to prevent oil from splashing onto the exhaust manifolds and other nearby parts. Start the engine and loosen the rocker retaining nut until a clicking noise begins. Then slowly tighten the nut just until the clicking noise stops. From this point, slowly tighten the nut about 1/4 turn at a time the specified number of turns (usually about 1 1/2 turns total). Wait a few seconds between each 1/4 turn to allow the lifter to leak down. Turning the adjusting too much at one time or too far can cause piston-to-valve contact.

Task 14   Inspect and replace camshaft drives (includes checking gear wear and backlash, sprocket and chain wear, overhead cam drive sprockets, drive belts, belt tension, tensioners, and cam sensor components).

When the camshaft gear teeth mesh directly with the crankshaft gear teeth, gear backlash may be measured with a dial indicator positioned against one of the camshaft gear teeth. Rock the cam gear back and forth and note the maximum reading on the indicator.

Some engines equipped with a timing chain and sprockets are fitted with a hydraulic tensioner. The tensioner uses pressurized oil from the lubrication system to eliminate timing chain play. Some manufacturers recommend measuring the installed length of the tensioner to determine chain wear. If the tensioner length exceeds the manufacturer's specifications, replace the timing chain.

On many V-type camshafts in block engines, a timing mark on the crankshaft sprocket must be aligned with a timing mark on the camshaft sprocket before the camshaft sprocket and chain are installed. Timing chain stretch and wear may be measured on these engines with a socket and flex handle installed on one of the camshaft sprocket retaining bolts. Rock the camshaft sprocket back and forth without moving the crankshaft gear, and measure the movement on one of the chain link pins on the camshaft sprocket.

Engines fitted with a timing belt should have the belt replaced at the mileage intervals recommended by the engine manufacturer. During belt replacement, the crankshaft, camshaft, and idler or other sprockets should be inspected. Check the sprocket teeth for wear and damage. Check idler pulleys or sprockets for dry or loose bearings.

Task 15   Inspect and measure camshaft journals and lobes.

To check camshaft straightness, rest the camshaft outer bearing journals on V-blocks and position a dial indicator against the middle cam bearing journal. Rotate the camshaft to determine runout. If the camshaft is not straight, replace it.

To measure camshaft lobes, use a micrometer to measure from the highest point on the lobe to a point on the opposite side of the lobe. Record the measurement. Then measure the lobe again at a position 90° from the first measurement. Subtracting the second measurement from the first gives camshaft lobe lift. Replace the camshaft if lift is not within specifications.

Use a micrometer to measure the diameter of each camshaft journal. Measure this diameter in several locations. If the diameter is less than specified, replace the camshaft.

Camshaft lobe lift can also be measured with the camshaft still in the engine by removing the valve cover and mounting a dial indicator on the cylinder head. Position the dial indicator so that it contacts the pushrod tip (rocker arm removed) or rocker arm directly above the pushrod (rocker arm still in place). The dial indicator stem must be parallel with the pushrod. Crank the engine by hand and note the highest and lowest dial indicator readings. The difference between these two numbers is camshaft lobe lift.

Task 16   Inspect and measure camshaft bore for wear, damage, out-of-round, and alignment; repair or replace according to manufacturer's specifications.

Inspect the camshaft bearing bores on a camshaft-in-block for scoring or other damage. Minor nicks or burrs can be removed with a round file. If the bearing bores are severely damaged, the block should be replaced.

On overhead cam engines without removable bearing caps, check the bearing bores for damage. Minor nicks or burrs can be removed with a round file. If the camshaft is binding, or the bearing inserts are worn unevenly, use a straightedge to check bearing bore alignment. If the bores are out of alignment, the head is warped and should be straightened or replaced.

On overhead cam engines with removable camshaft bearing caps, the camshaft usually runs directly against the aluminum head. After removing the bearing caps and the cam, place a straightedge across the cam bearing surfaces to measure bearing alignment. Measure the clearance between the straightedge and each bearing bore to determine the bore alignment. When the camshaft bearing bores are improperly aligned, replace the cylinder head. Check the bearing surfaces for scoring and other damage. To measure bearing bore out-of-round, install the bearing caps and torque the retaining bolts to specifications. Then measure bore diameter at several locations around the bore using a telescoping gauge. If the bearing surfaces and bores are in good condition, use Plastigage to measure the bearing clearance.

Task 17   Time camshaft(s) to crankshaft.

With the timing belt or chain cover removed, camshaft timing may be checked by noting the positions of marks on the camshaft and crankshaft sprockets. These marks must be aligned as indicated in the vehicle manufacturer's service manual.

On many OHV pushrod engines, the crankshaft sprocket is installed on the crankshaft nose and the crankshaft is rotated to position piston #1 at TDC. At this point, a mark stamped onto the crankshaft sprocket is pointing directly upward (toward the camshaft). The camshaft sprocket is then temporarily bolted to the cam and used to rotate the cam until a mark stamped on the cam sprocket is pointing directly downward (toward the crankshaft). The sprocket is then removed from the cam (without allowing the cam to rotate). The timing chain is looped over the cam gear, the mark on the cam gear is positioned directly downward, and the chain is looped around the crankshaft sprocket. When the cam sprocket is attached to the cam, the timing marks on the crank and cam sprockets should be pointing toward one another.

Single overhead camshaft engines fitted with a timing belt often use a similar procedure. After positioning the crankshaft so that piston #1 is at TDC, the camshaft is rotated to align a mark on the cam sprocket with a mark on the cylinder head. The timing belt is then installed.

The procedure used to time camshafts on double overhead camshaft engines varies from manufacturer to manufacturer. On some engines the cam sprockets are friction fitted to the cams. On these engines the cams can be rotated after the timing belt is installed. When the cams are rotated to the proper positions, the bolts locking the cam sprockets to the cams are tightened. Other DOHC engines use a procedure similar to that of many SOHC engines.

Valve timing may be checked by observing the valve position in relation to the piston position. With any piston at top dead center (TDC) on the compression stroke, the intake and exhaust valves for that cylinder should be completely closed. When the piston is at TDC on the exhaust stroke, the intake valve should be opening, and the exhaust valve should be closing. This position is called valve overlap. If the valves do not open properly in relation to the crankshaft position, the valve timing is not right. Incorrect valve timing may cause low power or, in extreme cases, bent valves due to piston-to-valve contact.

Task 18   Reassemble and install cylinder heads and gaskets; replace and tighten fasteners according to manufacturers' procedures.

Clean and inspect the cylinder block deck in preparation for head installation. Remove any minor nicks or burrs using a file. Make sure that all head positioning dowels, if used, are in place in the block. Run a tap into cylinder head bolt threaded holes. Then used compressed air to eject any debris from the threaded holes. Always wear eye protection when using compressed air to clean surfaces or openings. Allowing debris or fluid to remain in threaded holes will cause false torque readings when the head bolts are tightened. Coolant or combustion leaks may result. If the holes are blind holes, fluid or debris at the bottom of the holes may cause the block to crack when the bolts are tightened.

Many newer engines are fitted with torque-to-yield (TTY) cylinder head bolts. These bolts are usually tightened to a specific torque and then rotated tighter a specified number of degrees. Torque-to-yield bolts are permanently stretched as they are tightened and produce a more uniform clamping force. Most, but not all, TTY bolts must replaced with new bolts once they are loosened. Check the manufacturer's service manual for information.

Most modern head gaskets are installed dry, without any type of sealer. When positioning a head gasket on the block, make sure that any orientation marks (up, front, left, right, etc.) are followed.

After double-checking the cylinder bores for tools, shop towels, dropped fasteners, etc., set the cylinder head on the engine block. Check the manufacturer's recommendation regarding thread lubricants or sealers. Bolts that are threaded into blind holes are often lubricated with a few drops of engine oil—some on the threads and some on the underside of the bolt head. Bolts that thread into the water jacket are often coated with a waterproof sealer.

Insert the head bolts in their holes and hand tighten them. Then tighten the bolts to specifications following the procedure and sequence specified by the engine manufacturer.

C. Engine Block Diagnosis and Repair (18 Questions)

Task 1   **Disassemble engine block and clean and prepare components for inspection.**

Mount the engine on a stand and remove the oil pan drain plug. Allow any oil that has accumulated in the pan during engine removal to drain into a pan. Remove the lifters from the block and store them in an organizer for later inspection. Turn the engine block upside down, remove the oil pan bolts, and remove the pan from the block. If the pan is "glued" to the block with RTV sealant, strike a strong corner of the pan with a rubber mallet to loosen it.

If an oil pump or balance shaft assembly is mounted to the bottom of the block inside the oil pan, unbolt and remove it from the block. On chain-driven balance shaft mechanisms, the timing cover may have to be removed to access the drive sprocket and chain.

Check the crankshaft and connecting rod bearing caps to see if they are marked for position and direction. Main bearing caps often have numbers and arrows cast into each cap. Arrows typically point to the front (timing device end) of the engine. Connecting rods are often stamped with the cylinder number on both the rod body and cap, near the cap parting line. Take note of the direction that the numbers (or rod oil squirt holes) point for all connecting rods. If rod or main caps are not marked, mark each one with a number punch, center punch, or scratch awl.

Inspect the top of the cylinder bores for ring ridges. If the ring ridge is severe, it should be removed *before* attempting to remove the piston/connecting rod assemblies. Use a ridge reamer to remove the ridge (see Task C.4).

Loosen the connecting rod cap bolts or nuts and remove the caps. Keep used bearing inserts with their caps for later inspection. If bolts are held captive in the connecting rod bodies, place a short length of fuel line hose over each bolt to protect the crankshaft journals during piston/rod removal.

Carefully push each piston/connecting rod assembly out of its cylinder. When an assembly is removed, immediately reinstall its mating rod cap and nuts.

Remove the harmonic balancer or pulley hub bolt from the crankshaft nose and remove the balancer or hub. Some simply slide off the nose, but most are a press fit. Use a special harmonic balancer removal/installation tool to remove a press-fit balancer. Using a jawed puller to remove a press-fit balancer will permanently damage the balancer.

Remove the timing chain/belt cover bolts and remove the cover. On engines fitting with timing chain tensioning devices, compress and lock the tensioner shoe in place, if possible. Remove the tensioner and any timing chain guides from the front of the block. Remove the oil slinger, if one is present, from the crankshaft nose.

On camshaft-in-block engines, unbolt and remove a bolted-on camshaft sprocket along with the timing chain. Remove the camshaft thrust plate bolts and the thrust plate, if one is present. If the engine has a balance shaft mounted in the engine "V," remove the shaft drive mechanism.

On engines fitted with a radial-type crankshaft rear main oil seal, pry the seal out of its bore or unbolt the seal mounting plate and seal from the back of the engine. Remove the crankshaft main bearing cap bolts and the bearing caps. Lift the crankshaft out of the block.

Carefully withdraw the camshaft from the cylinder block, being careful to avoid nicking the bearing bores or the camshaft lobes. Use a camshaft bearing installer/remover tool to drive the bearings from their bores.

If the engine has block-mounted balance shafts, remove them now, following the engine manufacturer's instructions.

Knock any core or freeze-out plugs loose by striking one edge with a blunt chisel. Do not strike the plug too hard and do not try to drive the plug straight into the water jacket. This could cause a bulge in a cylinder wall if the plug is driven into the wall. When the core plug tilts in its bore, grab an edge of the plug with a pliers and pull the plug out of the block. Remove all oil gallery plugs.

Engine parts can be cleaned several different ways. Iron or steel parts can be soaked in a tank full of a heated alkaline solution (i.e., hot-tanked). This will remove oil, sludge, hard, baked-on carbon deposits, and mineral deposits in the coolant passages. Never put any aluminum parts in a hot tank—the caustic solution will corrode aluminum.

Many shops have cold solvent parts washers. Small to medium size parts can be placed in the washer and sprayed with a stream of mineral oil based solvent. Brushes or scrapers can be used to remove stubborn deposits from the parts.

Engine parts can also be cleaned in a thermal cleaner. These are actually large ovens that heat parts to temperatures between 650 and 800°F (343 and 427°C) to oxidize the contaminants. After the thermal cleaning process, the ash is removed by shot blasting or washing the parts. The temperature inside a thermal cleaner can also be reduced to clean aluminum parts without damaging them.

Regardless of which cleaning method is used, always perform a careful inspection of oil passages in the cylinder block, head, crankshaft, and all other parts. This is especially important if the engine suffered a major failure like a spun bearing or a severely worn camshaft and lifters. Metal particles will become lodged in the oil passages and can be difficult to remove. Rod out all small diameter passages to make sure that they are unobstructed. Use a long, slender rifle brush to thoroughly clean oil galleries.

Task 2   Visually inspect engine block for cracks, corrosion, passage condition, core and gallery plug holes, and surface warpage; determine necessary action.

A cast-iron block may be inspected for cracks with a electromagnetic crack detector. A dye penetrant may be used to check for cracks in an aluminum block. If the engine was consuming coolant, it may be necessary to pressurize the block with a pressure tester to check for cracks.

The cylinder head mounting surface (or "deck") on the block must be checked for warpage with a straightedge and a feeler gauge. First, check for minor nicks and burrs that may affect the warpage measurements. Minor nicks and burrs may be removed using a whetstone or a file. If deck warpage exceeds manufacturer's specifications, the block surfaces must be resurfaced. If warpage limits are not available, a general rule is that warpage under 0.005 inch (0.127 mm) is acceptable. Bolting a flat head to a warped block may cause head gasket or valve seat failures.

Task 3   Inspect and repair damaged threads where allowed; install core and gallery plugs.

Inspect all threaded bolt holes in the cylinder block, especially the cylinder head bolt holes. Check holes for sludge or debris that may have been missed during cylinder block cleaning. Run the appropriate size tap into each of the cylinder head bolt threaded holes to make sure that the threads are clean. Then put a few drops of oil on a head bolt and thread it into each hole by hand.

If the threads in a bolt hole are damaged, the hole may be drilled to a larger size and rethreaded. Then a thread repair insert or heli-coil can be installed in the oversize hole. The result is a threaded hole that is the same size as the original. Different types and brands of thread inserts are available. In most cases, however, the insert is threaded onto a special installation tool, coated with thread locking compound, and then threaded into the oversize hole. The installation tool is then removed and a hammer and punch are used to break off a tang at the bottom of the insert.

Inspect oil gallery plug threaded holes for dirt or damage. Do not run a tap very far into these holes since they usually have tapered pipe threads. Run a long brush (called a rifle brush) down the oil galleries to make sure that all debris has been removed. Then coat the new oil gallery plugs with teflon tape or an oil-resistant sealer and thread them into the block. Do not overtighten the plugs. If there are small core plugs at the ends of the oil galleries, coat the edges of new plugs with an oil-resistant sealer and then drive the plugs into the block. Use a cold chisel and hammer to cross stake the end of the bores after the core plugs are installed.

Clean the freeze out plug bores with emery cloth before installing new plugs. If a bore is damaged, it may be repaired by boring it to the next specified oversized plug. Oversized core plugs are stamped with the letters OS. Before installing a new plug, coat the sealing edge with a nonhardening, water-resistant sealer. Drive the plugs into the block using the proper special driving tool. Make sure that the plug goes into the bore squarely to prevent leaks.

Task 4   Inspect and measure cylinder walls; remove cylinder wall ridges; hone and clean cylinder walls; determine need for further action.

If the ring ridge at the top of each cylinder has not already been removed, remove it now using a ridge reamer. While using the ridge reamer, be careful to avoid marking the cylinder walls below the ring ridge. Do not remove any metal from the cylinder wall below the ring ridge. Failure to remove the ring ridge may cause piston ring lands and/or the top compression ring to crack or break after the engine is assembled and started.

Use a dial bore gauge to measure the cylinder diameter in three vertical locations. These locations are just below the ring ridge at the top of the cylinder, in the center of the ring travel, and just above the lowest part of the ring travel. Cylinder taper is the difference in the cylinder diameter at the top of the ring travel compared to the diameter at the bottom of the ring travel.

In each of the three vertical cylinder measurement locations, measure the cylinder diameter in the thrust direction and in the axial direction. Cylinder out-of-round is the difference between the cylinder diameter in the thrust and axial directions. If cylinder out-of-round exceeds specifications, rebore the cylinder.

If cylinder wear, out-of-round, and taper are within specifications, the cylinders may be deglazed. Very mildly worn cylinders should be deglazed with a brush hone, which removes material very slowly. Moderately worn cylinders (still within specifications) may be deglazed with 220 or 280 grit stones installed on a cylinder hone. When the honing operation is completed, the cylinders should have a 50° to 60° crosshatch pattern. After deglazing, the cylinder should be cleaned with hot, soapy water and a stiff-bristle brush.

Ordinary solvent will *not* remove grit from pores in the cylinder wall—use hot soapy water. The bores are clean when a clean, lint-free cloth is used to wipe them and the cloth does not get dirty. When the bores are clean, rinse the block and dry it thoroughly. Coat all machined surfaces with a light coating of the manufacturer's recommended engine oil.

If one cylinder requires reboring, most manufacturers recommend reboring all the cylinders to the same size. Cylinder reboring usually is done with a specialized piece of equipment called a boring bar. Cylinders must be honed after boring. A honing machine is usually used for this. After cylinder honing, the same procedure for block cleaning should be followed as previously discussed in cylinder deglazing.

Task 5   Visually inspect crankshaft for surface cracks and journal damage; check oil passage condition; measure journal wear; check crankshaft sensor reluctor ring (where applicable); determine necessary action.

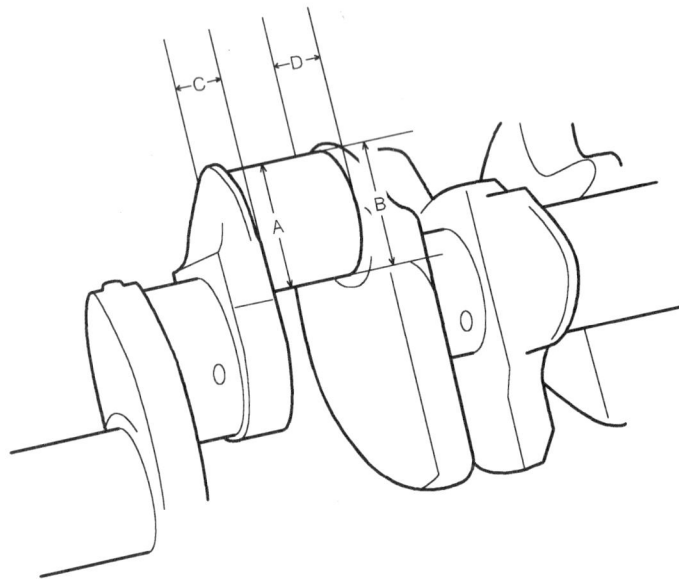

Referring to the figure above, use a micrometer to measure each crankshaft journal for vertical taper (difference between measurement A and measurement B), horizontal taper (difference between measurement C and measurement D), and out-of-round (difference between measurement A and measurement C or a difference between measurement B and measurement D). The out-of-round should be measured at two locations on each side of the journal. If the journal is out-of-round, the taper exceeds specifications, or journal scoring is evident, journal grinding is required.

Check the old main bearing inserts for uneven wear. If one or two bearings are worn more than the others, the crankshaft may be bent (or the bearing bores in the block may be misaligned). Check for a bent crankshaft by measuring main journal runout using V-blocks and a dial indicator. A bent crankshaft should be straightened and reground or replaced.

Run a slender rod through all oil passages drilled into the crankshaft to check for obstructions. Check the crankshaft for cracks using an electromagnetic crack detector.

Task 6   Inspect and measure main bearing bores and cap alignment and fit.

Check main bearing bore alignment before bearing caps are installed. The block should be resting on a flat surface, *not* hanging from an engine stand by the flywheel end. Lay a straightedge across the bores and check alignment using feeler gauges. If bearing bores are not aligned, the block can be line bored.

Check the bearing cap and cylinder block mating surfaces for nicks and burrs. These can be removed using a file.

When measuring main bearing bore diameters, the bearing caps must be installed and the bolts properly torqued. Check bore diameter in three directions. The vertical measurement should not be larger than any of the others. A larger vertical reading indicates the bore is stretched. Out-of-round measurements less than 0.001 inch (0.025 mm) are acceptable, provided that the vertical reading is not the largest.

Improper bore alignment and bore dimensions can be corrected by line boring. This operation, performed by machine shops, involves removing the bearing caps and planing a small amount of material from the cap surface that mates with the cylinder block. The caps are then reinstalled and torqued to specifications. A specialized piece of equipment called a line hone is used to "true up" the main bearing bores to their original diameters.

Task 7   Install main bearings and crankshaft; check bearing clearances and end play; replace/retorque bolts according to manufacturers' procedures.

Clean the main bearing bores in the cylinder block and bearing caps with solvent and allow the bore surfaces to dry. Do not oil the bores. Handle new bearing inserts carefully—avoid touching the bearing surface with your fingers. Wipe the back of the bearing inserts with a solvent-dampened cloth and allow the inserts to dry. Install the bearing inserts in the cylinder block and main bearing cap bores. The upper bearing halves are usually grooved and each contains an oil supply hole. Make sure that the oil hole in the bearing aligns with the oil hole in the bearing bore. Make sure that the tab on each bearing insert fits tightly in its bearing bore notch.

Carefully lay the crankshaft in the cylinder block.

To measure bearing clearance, install a strip of Plastigage across each journal. Then install the bearing caps and bolts, tightening the bearing cap bolts to the specified torque. Remove the bearing cap bolts and the bearing caps. Compare the width of the crushed Plastigage strip on the bearing journal to the scale provided on the Plastigage package to determine the bearing clearance.

Crankshaft end play may be measured by inserting a feeler gauge between the crankshaft thrust journal and the thrust lip on one of the main bearings. On some engines, a dial indicator is used to measure crankshaft end play while moving the crankshaft with a pry bar. Excessive end play may cause premature bearing wear or noise as the crank chucks back and forth in the block.

If bearing clearance and end play are within specifications, remove the main bearing caps and the crankshaft. Install a rope-type or split lip seal in its grooves in the block and rear main bearing cap. Oil the main bearing inserts, lay the crankshaft in the block, and install the main bearing caps.

Install the bearing cap bolts and torque them to specifications. Some engines are fitted with TTY main bearing cap bolts that must be replaced after being loosened. Check the manufacturer's service manual to determine whether the engine you are servicing has TTY main bearing cap bolts.

Task 8   Inspect camshaft bearings for unusual wear; remove and replace camshaft bearings; install camshaft, timing chain, and gears; check end play.

Inspect the camshaft bearings for scoring, roughness, and wear. Camshaft bearings or bearing bores should be measured at two different locations with a telescoping gauge. Measure the camshaft journals with a micrometer, and subtract the journal diameter from the bearing diameter to obtain the clearance. If the wear exceeds specifications, replace the bearings.

The type of tool needed to remove and install the camshaft bearings depends upon the engine design. Most overhead valve (OHV), camshaft-in-block engines will use a camshaft bushing driver and hammer. The right size mandrel is selected to fit the bearing. Turning the handle tightens the mandrel against the camshaft bearing. Then the bearing is driven out by hammer blows. The same tool is used to replace the bearings. Some overhead camshaft (OHC) engines require a special puller/installer.

Never attempt to remove and install camshaft bearings in an OHC cylinder head using a bushing driver. The bearing supports may be bent or broken due to the hammer blows, especially on aluminum heads.

When installing cam bearings, it is very important that the bearing insert be properly positioned in the bearing bore. Be absolutely sure that any oil hole(s) in the bearing insert align with oil supply passages in the bearing bore. This may mean that the insert is positioned toward the front of the bore, the back of the bore, or even the center of the bore. The position is not important so long as the oil holes line up.

Many overhead cam engines do not have removable camshaft bearings—the camshaft journals run directly against bearing bores machined into the aluminum cylinder head. Inspect the bearing surfaces on these engines for scoring, roughness, and wear. Measure the bearing bores at two different locations with a telescoping gauge. Measure the camshaft journals with a micrometer, and subtract the journal diameter from the bearing diameter to obtain the clearance. If the wear exceeds specifications, replace the cylinder head.

When the camshaft bearing bores are machined into the cylinder head, the bearing caps should be removed and a straightedge positioned across all the bearing bores. Insert a feeler gauge between the straightedge and each bearing bore to measure any misalignment. Misalignment indicates that the cylinder head may be warped. Have a machine shop check for this. Cylinder heads can sometimes be straightened. Severely warped heads should be replaced.

Task 9 Inspect auxiliary (balance, intermediate, idler, counterbalance, or silencer) shaft(s) and support bearings for damage and wear; determine necessary action.

Balance shafts are found on many 4- and 6-cylinder engines. Some rotate at crankshaft speed; some rotate at twice crankshaft speed. Removal and installation procedures vary widely; refer to the manufacturer's service manual for service information.

Some balance shafts are mounted to the bottom of the cylinder block and are chain driven off the crankshaft. Some are gear driven by a large gear machined into a disc that is part of the crankshaft. Belt-driven balance shafts are often mounted inside the bottom of the cylinder block, much like camshafts on camshaft-in-block engines. On some popular V-type engines, a balance shaft is mounted in the cylinder block directly above the camshaft and gear driven off the camshaft.

Once removed, balance shafts should be checked for runout with the same procedure used for measuring camshaft runout. The balance shaft journals should be measured for taper with the same procedure for measuring crankshaft journal taper. When the balance shafts are installed, they must be properly timed to the crankshaft or severe engine vibration may occur upon engine startup.

Task 10 Inspect, measure, service, repair, or replace pistons, piston pins, and pin bushings; identify piston and bearing wear patterns that indicate connecting rod alignment problems; determine necessary action.

Inspect pistons for cracks and damage from overheating. Pistons with these conditions should be replaced. Inspect the piston skirts for uneven wear. Wear on the edges of the piston skirt next to the wrist pin hole may be caused by a bent or twisted connecting rod.

Clean the piston ring grooves using a ring groove cleaning tool. Be careful to avoid removing material from the bottom of the groove. Make sure that oil drain holes at the bottom of the oil ring groove are not obstructed.

To measure piston ring side clearance, insert a new ring in the piston groove and position a feeler gauge between the ring and the groove. If ring side clearance is excessive, the piston should be replaced.

If the piston passes the inspections mentioned previously, check for a worn wrist pin bushing. Clamp the connecting rod body lightly in a vise (use soft jaw covers) and attempt to rock the piston against the connecting rod sideways (at a right angle to normal piston/connecting rod motion). If any play is noticeable, the piston wrist pin bore, wrist pin, or connecting rod small end bushing are worn and the piston and connecting rod must be separated. If no play is noticeable and the piston is to be reused, the components need not be separated.

Pistons must be fitted to their cylinder bores. If piston-to-cylinder wall clearance is excessive, piston slap may be noticeable. If there is not enough clearance, piston scuffing will occur. Piston clearance is typically 0.001 to 0.002 inch. Check piston clearance by measuring the cylinder bores and the piston diameters.

To measure piston diameter, position a micrometer to contact the piston thrust surfaces (the surfaces at right angles to the piston, or wrist, pin bore). The exact measuring location varies according to manufacturer. Most manufacturers, however, specify a point about 3/4 inch below the wrist pin centerline. Slightly undersize pistons may be knurled and reinstalled if they are in otherwise good condition. Severely worn pistons must be replaced.

Task 11   Inspect connecting rods for damage, alignment, bore condition, and pin fit; determine necessary action.

Inspect the connecting rods for cracks and obvious damage. Remove the cap nuts, caps, and bearing inserts. Check that the cap bolts are not loose in the rod body.

Inspect the bearing inserts for uneven wear. If the front and rear edges of a bearing are worn more than the center area, the rod may be bent or twisted. If the bearing inserts are worn more at the parting line areas, the rod big end bore may be stretched. Machine shops have special jigs used to check for bent/twisted connecting rods. If rod bend or twist exceeds specifications, replace the connecting rod.

Measure the connecting rod big end bore for taper, out-of-round, and proper bore size. If any of these dimensions not within specifications, the connecting rod should be rebuilt or replaced.

On piston/connecting rod assemblies with free-floating wrist pins, remove the pin retainer circlips or snaprings and slide the wrist pin out of its bores. Take note of piston to rod orientation and separate the piston and rod. On piston/connecting assemblies with press-fit wrist pins, use the appropriate press and adapters to remove the wrist pin. Measure the connecting rod small end bore (or "eye") diameter. If bore diameter exceeds specifications, it may be possible to ream out the bore and install an oversize wrist pin. Some rod eyes have a pressed in bushing. If the bore is worn on this type of rod a new bushing can be installed. The new bushing must be reamed to fit the wrist pin.

If the rod beam has an oil squirt hole, make sure that the passage from the hole to the big end bore is not obstructed.

Task 12   Inspect, measure, and install or replace piston rings; assemble piston and connecting rod; install piston/rod assembly; check bearing clearance and side play; replace/retorque fasteners according to manufacturers' procedures.

When new piston rings are installed, ring end gap must be checked and adjusted. Compress a ring just enough so it fits in a cylinder. Insert a piston in the cylinder upside down (crown end first) and push the ring down near the bottom of the cylinder. The ring should be about _" from the bottom. Use a feeler gauge to measure the gap where the piston ring ends meet. If the gap is too small, the ring ends must be filed to increase the gap. If the gap is too large, the wrong rings were selected or the cylinder was bored/honed incorrectly.

If the pistons and connecting rods were separated, reassemble them. Position the rod eye inside the piston, making sure that the parts are oriented correctly. Most pistons have a notch or arrow in their crown that must point toward the front of the engine. Connecting rod orientation varies according to the engine manufacturer. Refer to notes made during disassembly or check the service manual for information. On engines with free-floating wrist pins, dip the pin in engine oil and install it in its bore. Install new circlips or snaprings. When installing press-fit wrist pins, many manufacturers recommend that the rod eye be heated before pin installation. Some manufacturers recommend that the piston be heated in a piston heater, as well. When components are heated as necessary, use a press and the appropriate adapters to install the pin.

When installing the piston rings on the piston, install the oil rings first. Follow instructions that come with the rings to position the upper and lower rail gaps with respect to the expander gap. Oil ring rails can be spiraled into their slots. When installing compression rings, make sure that any marks stamped into the ring are facing upward. Install the bottom compression ring first and then the top compression ring using a piston ring expander. Do *not* spiral compression rings onto the piston. This can bend the rings, causing them to resemble a lock washer.

Clean the bearing bores in the connecting rods and rod caps with solvent and allow the bore surfaces to dry. Do not oil the bores. Handle new bearing inserts carefully— avoid touching the bearing surface with your fingers. Wipe the back of the bearing inserts with a solvent-dampened cloth and allow the inserts to dry. Install the bearing inserts in the rod and cap bores. If the rod has an oil squirt hole, make sure that the oil hole in the bearing aligns with the oil hole in the rod body. Make sure that the tab on each bearing insert fits tightly in its bearing bore notch.

Before installing the piston/connecting rod assembly in the block, install short lengths of fuel line over the rod cap bolts and position the piston ring end gaps according to the manufacturer's instructions. If no instructions are given, it is common practice to position the gaps as follows:

- Oil ring expander gap facing the front of the engine, directly above the wrist pin centerline.
- Upper oil ring rail gap 45° to one side of the expander gap.
- Lower oil ring rail gap 45° to the other side of the expander gap.
- Bottom compression ring gap on the left side of the piston (90° from the wrist pin).
- Top compression ring gap on the right side of the piston (90° from the wrist pin).

Rotate the crankshaft to position the crank pin for the piston/rod being installed at BDC, install a ring compressor on the piston, and slide the piston/rod into the block. Push the rod body against the crank pin, remove the rubber protective hose, and temporarily install the rod cap and nuts. Install the remaining piston/rod assemblies.

To measure bearing clearance, remove the rod cap and install a strip of Plastigage across the crankshaft journal. Then install the bearing cap and nuts, tightening the nuts to the specified torque. Remove the bearing cap nuts and the bearing caps. Compare the width of the crushed Plastigage strip on the bearing journal to the scale provided on the Plastigage package to determine the bearing clearance.

Measure the side clearance at each connecting rod by inserting a feeler gauge between the side of the connecting rod and the edge of the crankshaft journal. If side clearance exceeds specifications, the sides of the connecting rod or crankshaft journal are worn.

Task 13   Inspect, reinstall, or replace crankshaft vibration damper (harmonic balancer).

A special puller and installer tool are required to remove and install the vibration damper. Using a regular gear puller to remove the vibration damper will damage the damper.

Inspect the rubber between the inner hub and outer inertia ring on the vibration damper. If this rubber is cracked, oil-soaked, deteriorated, or protruding from the damper, replace the damper. If the inertia ring on the damper is loose or has shifted forward or rearward on the hub, replace the damper. Inspect the vibration damper hub for cracks or a damaged keyway. If either of these conditions is present, replace the damper. Inspect the seal contact area on the hub for a wear groove or scoring. If either of these conditions is present, replace the damper or install a sleeve on the hub to provide a new seal contact area.

Task 14 Inspect crankshaft flange and flywheel mating surfaces; inspect and replace crankshaft pilot bearing/bushing (if applicable); inspect flywheel/flexplate for cracks and wear (includes flywheel ring gear); measure flywheel runout; determine necessary action.

Inspect the crankshaft flange and the flywheel-to-crankshaft mating surface for metal burrs. Remove any metal burrs with fine emery paper. Be sure the threads in the crankshaft flange are in satisfactory condition. Replace the flywheel bolts and retainer (if fitted) if any damage is visible on these components. Install the flywheel, retainer, and bolts, and tighten the bolts following the torque and sequence provided by the engine manufacturer.

Inspect the flywheel for scoring and cracks in the clutch contact area. Minor score marks and ridges may be removed by resurfacing the flywheel. If deep cracks or grooves are present, the flywheel should be replaced.

Mount a dial indicator on the engine block or flywheel housing and position the dial indicator stem against the clutch contact area on the flywheel. Rotate the flywheel to measure the flywheel runout. If runout exceeds specifications, replace the flywheel.

Insert a finger in the inner pilot bearing race and rotate the race. If the bearing feels rough or loose, replace the bearing. Check a pilot bushing to verify that it is not loose. A transmission input shaft may be positioned in the pilot bushing to check for excessive play. If too much play exists, replace the bushing. A special puller may be used to remove the pilot bearing or bushing. The proper driver must be used to install the pilot bearing or bushing. Always verify that the transmission input shaft fits in a new bushing before attempting to install the engine (or transmission).

Inspect the starter ring gear for excessive wear or damage. On manual transmission flywheels, the starter ring gear is often replaceable. Remove the old gear by drilling a hole through the gear at the "root" between two teeth. Then position a cold chisel between the two teeth and strike it with a hammer. Take note of whether the gear has a chamfer on one side before removing it. To install the new gear, first heat it to about 400°F in an oven. Then slip it over the flywheel body and allow it to cool.

Task 15 Inspect and replace pans, covers, gaskets, and seals.

Always inspect the gasket mounting surfaces on sheet metal pans for warpage, and look for dished retaining bolt holes. Dished mounting holes and warped mounting surfaces must be straightened by hammering them flat again. Use a straightedge to verify that gasket surfaces are flat.

Gaskets are used to seal minor variations between two flat surfaces. Oil pan gaskets or rocker arm cover gaskets are usually manufactured from cork, rubber, or a combination of rubber and cork or rubber and silicone. Timing cover, water pump, and thermostat housing gaskets on older engines were made of specially treated paper. On newer engines, these same components are sealed to the block using synthetic rubber O-rings or silicone "spaghetti" seals. Paper gaskets should be coated with a nonhardening sealer prior to installation. Synthetic rubber and silicone seals are often installed without coatings or cements.

Lip type seals are classified as springless or spring loaded. Springless seals are used in front wheel hubs where they seal a heavy lubricant. A spring-loaded seal contains a garter spring to force the seal lip against a rotating shaft. The spring allows the seal to compensate for lip wear, shaft runout, and bore eccentricity. Some seal lips are specially formed to oscillate during operation. This actually "pumps" oil back into a housing. The metal body of many seals is painted with a sealer to prevent oil from leaking between the seal body and its mounting bore. When installing a seal, make sure that the garter spring faces toward the fluid being sealed.

Before installing a seal, inspect the shaft and seal bore for scratches and roughness. Seal lips should be lubricated before installation, and the proper seal driver must be used to install the seal. Be sure that a seal is driven squarely into its bore and is driven to the proper depth.

Task 16   Assemble engine parts using formed-in-place (tube-applied) sealants or gaskets.

Room-temperature vulcanizing (RTV) sealer may be used in place of conventional gaskets on oil pans and rocker arm covers. RTV sealer dries in the presence of air by absorbing moisture from the air.

All old gasket material must be removed from the surfaces being sealed before RTV sealer is applied. Denatured alcohol or a chlorinated solvent must be used to clean the RTV sealer contact area. If oil-based solvents are used to clean the mounting area, an oily residue is left on the area that prevents RTV sealer adhesion. Apply a 1/8-inch (3.2-mm) diameter bead of RTV sealer to the center of one surface to be sealed. This bead must surround bolt holes in the mounting area. Do not apply too much RTV sealer, and always assemble the components within five minutes after the RTV application or curing will occur. Fumes from high volatility RTVs can damage oxygen sensors, so be sure to use a sealer labeled "safe for O2 sensors" when installing an intake manifold.

Anaerobic sealer may be used in place of a gasket when two smooth, very flat surfaces are being joined. This sealer dries in the absence of air. Before applying anaerobic sealer, clean both surfaces being sealed following the same procedure used for RTV sealer.

D. Lubrication and Cooling Systems Diagnosis and Repair (9 Questions)

Task 1   Perform oil pressure tests; determine necessary action.

When testing engine oil pressure, the oil pressure switch or sending unit usually is removed from its bore and an oil pressure gauge is installed in its place. Oil pressure should be checked when the engine is at normal operating temperature, if possible. Check oil pressure at idle speed and a higher speed, such as 2,500 revolutions per minute (rpm).

Common causes of low oil pressure include worn camshaft or crankshaft bearings, a clogged or damaged oil pickup screen or tube, and a weak (or jammed) oil pump pressure regulator spring. Restricted pushrod passages will not cause low oil pressure.

If crankshaft main or connecting rod bearings are worn enough to cause low oil pressure, hammering or thumping noises should be audible from the crankcase.

If oil pressure is okay at startup but it drops suddenly during engine operation, the oil pan may contain debris such as broken timing sprocket teeth (nylon), broken valve stem seals (umbrella type), or excess RTV sealer. The debris moves around in the oil pan until it is sucked against the sump screen, piece by piece. Remove the oil pan to check for this. The debris may not be flushed out during an oil change.

Task 2   Disassemble, inspect, measure, and repair oil pump (includes gears, rotors, housing, and pickup assembly), pressure relief devices, and pump drive; replace oil filter.

Inspect the oil pump pressure relief valve for sticking and wear. If this valve sticks in the closed position, oil pressure will be too high. A pressure relief valve stuck in the open position results in low oil pressure.

On rotor type oil pumps, measure the thickness of the inner and outer rotors with a micrometer. When this thickness is less than specified on either rotor, replace the rotors or the oil pump. The following oil pump measurements should be performed with a feeler gauge:

- Measure pump cover flatness with a feeler gauge positioned between a straightedge and the cover.
- Measure the clearance between the outer rotor and the housing.
- Measure the clearance between the inner and outer rotors with the rotors installed.
- Measure the clearance between the top of the rotors and a straightedge positioned across the top of the oil pump.

On gear type oil pumps, inspect the gears and housing for scoring and excessive wear. When reassembling a gear type pump be sure to align any match marks stamped onto the pump gears.

Task 3   Perform cooling system tests; determine necessary action.

A pressure tester may be connected to the radiator filler neck to check for cooling system leaks. Operate the tester pump and apply 15 psi (103 kPa absolute) to the cooling system. If the gauge pressure drops more than specified by the vehicle manufacturer, the cooling system has a leak. Inspect the radiator, hoses, gaskets, core plugs, water pump weep hole, etc. for external leaks while the cooling system is pressurized. If there are no visible external leaks, check the front floor mat for coolant dripping out of the heater core. If there are no external leaks, check the engine for internal leaks.

A cracked or porous cylinder block or cylinder head may allow coolant to seep into the crankcase. This may cause the engine oil level to rise over time. It may also cause sludge to form. If a cylinder block or head crack enters the combustion chamber, combustion gases are usually forced into the cooling system. To check for this, connect the pressure tester to the radiator filler neck and start the engine. Cooling system pressure will rise immediately upon engine startup and coolant will be forced into the recovery system reservoir.

Minor cracks may not cause a dramatic rise in cooling system pressure. To check for minor combustion chamber leaks, use a special leak detector kit. The kit usually consists of a colored fluid that reacts to hydrocarbons and a clear cylinder that fits into the radiator filler neck. Some test fluid is poured into the clear cylinder and the cylinder is placed on the radiator filler neck. The engine is started and a bulb on top of the cylinder is squeezed to pull air from the cooling system through the test fluid. If the fluid changes color (from blue to yellow), combustion gases are entering the cooling system.

When fitted with the proper adapter, a cooling system pressure tester may also be used to check the radiator cap. When the tester pump is operated, the cap should hold the rated pressure.

Task 4   Inspect, replace, and adjust drive belts, tensioners, and pulleys.

Inspect the accessory drive belts for condition and tension. On conventional V-belts, the sides of the belt are the friction surfaces, so check the sides for cracks, glazing or loose cord material. Replace a belt showing any of these conditions. If a V-belt is severely worn, it may contact the bottom of the pulleys. Replace severely worn belts. If a belt is severely worn on just one side, check pulley alignment. V-belt pulleys must be aligned within 1/16 inch (1.6 mm) per foot of belt span. If pulleys are not aligned, check for loose accessory mounting bolts, missing spacers, or bent brackets.

V-belt tension can be checked using a variety of special testers. With one type of tester, the tool is placed over the belt at the center of a belt span. Squeezing the tool handles causes the tool dial to display belt tension, usually in pounds. Belt tension can also be checked by measuring the amount of belt deflection with a ruler. Use your thumb to press on the belt at the middle of a span while holding the ruler next to the belt. If belt tension is correct, the belt should deflect 1/2 inch (12.7 mm) for every foot (30.5 cm) of belt span.

A moderately loose or worn belt may cause a squealing noise when the engine is accelerated. A severely worn or loose belt may cause a discharged battery, engine overheating, or a lack of power steering assist. An overtightened belt may fail suddenly, or damage the alternator front bearing. An overtightened belt can also cause the upper half of the crankshaft front main bearing to wear prematurely.

When repositioning an accessory device (alternator, power steering pump, etc.) to adjust belt tension, always look for the pry points provided by the manufacturer. Some devices have slots for inserting a large screwdriver or pry bar. Others have built-in square holes to accommodate a 1/2 inch breaker bar or ratchet. Never pry on a power steering pump housing to tighten the drive belt. These housings are not meant to withstand such abuse and will be damaged, possibly causing a fluid leak.

Inspect serpentine or V-ribbed belts, looking for missing sections of ribs, severe rib cracking, or a damaged belt backside. These conditions call for immediate belt replacement.

Serpentine or V-ribbed belts must also be properly tensioned, but these belts are usually fitted with automatic tensioners. The tensioner automatically adjusts belt tension as the V-ribbed belt stretches. These tensioners often have built-in wear indicator scales. So long as an arrow on the tensioner is located between two lines, the belt is not excessively stretched. When the arrow moves outside the lines, the belt must be replaced.

Task 5   Inspect and replace engine cooling and heater system hoses.

Check all cooling system hoses for loose clamps, leaks, and damage. Look for cracks, abrasions, bulges, and swelling. Check for hard spots due to heat damage from close proximity to exhaust system components. Also look for shiny spots caused by contact with accessory mounting brackets or other components. These spots may indicate weak spots that could cause a hose to burst. Check the hoses for soft or gummy areas due to contact with engine oil, power steering fluid, or automatic transmission fluid.

Squeeze each hose along its entire length to check for hard or soft areas. Also listen for crackling or crunching noises while squeezing which would indicate that the reinforcing fabric is faulty or the inner liner has deteriorated. Lower radiator hoses often contain a steel spring to prevent the hose from collapsing, so you may not be able to perform the squeeze test.

When in doubt about a hose's condition, remove it and inspect the inner liner. If the liner is cracked or otherwise deteriorated, replace the hose.

Be careful when removing a faulty hose. Aggressive twisting and pulling can damage a heater core or radiator tank. If the hose is stuck to the fitting, slit the end of the hose to make removal easier.

When installing a new hose, make sure that it fits properly. Avoid twisting or stretching the hose. A hose that is too short may fail when the engine shifts during acceleration.

Task 6   Inspect, test, and replace thermostat, bypass, and housing.

The thermostat may be tested after it has been removed from the engine. Submerge the thermostat in a pan of water and put a thermometer in the water. Suspend the thermostat and the thermometer above the bottom of the pan. Allowing them to lay on the bottom of the pan will not allow an accurate test. Heat the water while observing the thermostat valve and the thermometer. The thermostat valve should begin to open when the temperature on the thermometer is equal to the rated temperature stamped on the thermostat. Replace the thermostat if it does not open at the rated temperature.

Always replace a thermostat with one having the correct temperature rating. Do not install a "hotter" thermostat in an attempt to speed up engine warm-up time. The engine will warm up at the same rate, but operate at a higher temperature. Do not remove a functioning thermostat from an engine that is overheating. While the engine may stop overheating, coolant will flow through the engine too quickly to absorb heat adequately. Hot spots will develop in the cooling system, especially in the cylinder heads. A cracked head can result.

Be sure to install a thermostat in the correct direction and orientation. Many thermostats have an arrow indicating which way coolant should flow through the thermostat. Some thermostats have a vent hole, "jiggle" pin, or check ball assembly mounted toward the edge of the mounting flange. This device, which allows trapped air to pass through the thermostat, must be oriented properly. In most cases the device must point upward. Check the engine manufacturer's service manual for instructions.

Inspect the thermostat housing and bypass hose (if equipped) for cracks, deterioration, and restrictions. Thermostat housings are often made of sheet metal or a light alloy that corrodes rapidly when coolant is not changed at the recommended intervals. Replace a deteriorated thermostat housing or bypass hose.

Task 7   Inspect coolant; drain, flush, and refill cooling system with recommended coolant; bleed air as required.

The cooling system should be drained, flushed, and refilled with new (or recycled) coolant at the intervals specified by the vehicle manufacturer. This interval is often every two years. Coolant serves several important functions:
- Conducts heat away from hot engine parts.
- Prevents the cooling system from freezing.
- Prevents the cooling system from boiling.
- Lubricates the water pump seal.
- Inhibits corrosion.

If coolant is too old, too weak, or even too strong, it cannot perform all of these functions correctly.

A quick visual inspection will reveal obviously weak or contaminated coolant. If the coolant appears to be in good condition, try to determine how old it is. Coolant that is less than two years old may be left in the system if strength is acceptable. Check coolant strength (concentration) using a coolant hydrometer. Insert the hydrometer tube into the radiator filler neck, then squeeze and release the bulb on top of the tool. Read the scale on the hydrometer or count the number of floating plastic balls to determine coolant freezing point. Most vehicles are factory-filled with a 50/50 mixture of coolant and water. This provides protection against freezing down to a temperature of –34°F (–37°C).

Drain old, weak, or contaminated coolant from a cooling system by loosening the radiator cap, opening the radiator petcock, removing engine block drain plugs, and (if necessary) disconnecting the lower radiator hose. Engine coolant must be handled as a hazardous waste material or recycled. Coolant reconditioning machines are available to remove harmful particles and restore corrosion additives so the coolant can be returned to the cooling system.

When using a cooling system flush product to remove scum, scale, and rust deposits, follow the instructions provided with the flush. When using a two part acid/neutralizer flush, always use the neutralizer portion of the package.

If the radiator tubes and coolant passages in the block and cylinder head are restricted with rust and other contaminants, heavy-duty cooling system flushing equipment may be required. Always operate the flushing equipment according to the equipment manufacturer's directions, and be sure that your service procedure conforms to pollution laws in your state.

Refill the cooling system with a 50/50 mixture of coolant and water. This protects against freezing down to a temperature of –34°F (–37°C). Increasing the proportion of coolant will further lower the freezing point until the coolant/water ratio reaches about 70/30. At this point freezing temperature begins to rise. Another reason to avoid very strong coolant solutions or pure coolant is that coolant does not absorb heat as well as water, so cooling system efficiency is reduced.

On some vehicles, air pockets tend to develop in the cooling system as it is filled. If these air pockets are not bled off, engine overheating and even a cracked cylinder head can result. Some engines have a bleed fitting installed on the thermostat housing or an engine coolant passage to release trapped air. Loosen this fitting until all air is removed. On engines that do not have a bleed fitting, locate the highest point in the cooling system. If this point is a hose connection, loosen the hose to bleed off air. If the highest point is located at or near a coolant temperature switch or sensor, disconnect any wiring from the switch or sensor and loosen it just enough to allow air to escape. The bleeding procedure should be repeated after the engine has been operated for a few minutes.

Task 8   Inspect and replace water pump.

Check the water pump for leaking hose connections, mounting gaskets, and seals. Slow or hard-to-find leaks may be easier to find if a cooling system pressure tester is connected to the radiator filler neck.

Locate and examine the vent, or weep, hole in the water pump housing. The hole is usually in the underside of the housing, so use a small inspection mirror, if necessary. If the water pump seal is leaking, coolant will usually drip from the weep hole. A very slow leak may leave only coolant residue around the hole. Replace the pump if there is evidence of coolant at the weep hole.

If the customer complains about a growling noise coming from the engine, use a stethoscope to listen to the water pump bearing. A defective water pump bearing may cause a growling noise at idle speed. In some cases, the bearing starts to fail after being contaminated by coolant leaking past the pump seal.

With the engine shut off, grasp the fan blades or the water pump pulley, and try to move it from side to side. This will reveal any looseness in the water pump bearing. If there is any side-to-side movement in the bearing, the water pump should be replaced.

When replacing a pump, always compare the new pump to the old one. Two pumps may look very similar, but their impellers may rotate in opposite directions. In this case, the impeller blades will be shaped differently and installing the wrong pump will cause the engine to overheat.

On many engines, some of the water pump mounting bolts extend into the block water jacket. Be sure to use the specified sealant on these bolts or coolant may leak from the engine. The bolts may also seize in place, making future servicing difficult. Refer to the manufacturer's service manual for information to determine which bolts enter the water jacket.

Task 9 Inspect, test, and replace radiator, heater core, pressure cap, and coolant recovery system.

Examine the radiator for obvious damage or defects. Look for bent fins and fins clogged with dirt, road debris, or insects. These conditions greatly reduce radiator efficiency and can cause engine overheating. If damage is not severe, bent fins can usually be straightened using a special comb made for this purpose. Dirt and insects can be removed using a stream of low pressure water or compressed air.

Check the radiator for leaks or damp spots. Cracked solder seams and corroded tubes in copper/brass radiators can allow coolant to leak very slowly without leaving puddles of coolant. The same is true for aluminum/plastic radiators with cracked plastic tanks and leaking tank gaskets. Hard to find leaks can be located by removing the radiator, plugging the inlet and outlet fittings, and pressurizing the radiator with a cooling system tester. Submerge the radiator in a tank of water and check for bubbles.

If the engine is overheating, remove the radiator cap and look inside the radiator at the tubes, if possible. Check for clogged or restricted tubes. If you think the radiator tubes may be restricted, take the radiator to a specialty shop that has flow test equipment. This equipment can determine radiator flow capacity, which is often 20 gallons per minute or more. It is not usually possible for the typical general repair shop to achieve this flow rate.

Check the radiator cap for corrosion and damaged or deteriorated gaskets. Check the radiator filler neck seat, too. If the cap gasket or filler neck seat is damaged, the cooling system may not pressurize enough to prevent boilover. Coolant will be forced out of the cooling system and onto the ground or into the coolant recovery tank, if the vehicle has one. The engine may overheat.

If the vehicle is equipped with a coolant recovery system, check the gasket at the very top of the radiator cap. If this gasket is missing or leaking, coolant may be forced into the recovery tank when the engine warms up, but may not be drawn back into the radiator during cool down. Check the tube that connects the radiator filler neck to the reservoir for kinks, damage, or loose connections. These conditions may also allow coolant to flow into the tank, but not return to the radiator. Check the coolant reservoir for cracks, loose fittings and other damage. Some recovery systems use a reservoir cap that allows a length of tubing to hang down into the coolant. If this hose is missing or damaged, coolant cannot return to the radiator when the engine cools down.

Check the vacuum valve in the radiator cap. If the valve is stuck closed, a vacuum may form in the cooling system when the engine cools down. This vacuum may cause cooling system hoses to collapse. If the vacuum valve is stuck open, the system will not pressurize and coolant may be forced into the recovery tank. The pressure and vacuum valves in a radiator cap may be tested by fitting the proper adapter to a cooling system tester.

If the cooling system is losing coolant, check the heater core for leaks. Most leaks result in wet or damp floor mats or carpet in the passenger compartment. Coolant is easily identified by its green color, sweet smell, and slippery feel. Very small leaks in the heater core may not deposit coolant on the floor mats or carpet. Instead, the coolant is atomized and carried through the heater system ductwork. In these cases, ask the customer about excessive windshield fogging. Check the inside of the windshield for an oily or slippery residue.

Task 10   Clean, inspect, test, and replace fan (both electrical and mechanical), fan clutch, fan shroud, air dams, and cooling related temperature sensors.

On rear-wheel-drive vehicles, the engine cooling fan is usually mounted to the water pump shaft and belt driven off the crankshaft. Plain, direct-drive fan blade assemblies should be checked for loose mounting bolts, cracked blades, and loose rivets (if fan blades are riveted to a hub). A fan assembly that has any cracks should be replaced immediately. Fans with temperature sensitive clutches should be checked for bad bearings, leaking fluid, and seized or free-wheeling clutches. With the engine off, try to spin the fan by hand. It should spin smoothly with some resistance. If the bearing feels rough, or the fan spins without resistance, replace the clutch assembly. Grasp the fan blades and try to rock the fan from side to side. Too much play indicates a bad bearing and the clutch should be replaced. Check the bimetal coil on the front of the clutch. If it is wet or covered with dirt and grime, silicone fluid is leaking out of the fluid reservoir and the clutch should be replaced. To check for a seized fan clutch, start the engine and observe fan speed. When the engine is cold, the fan should not pull much air through the radiator, even when the engine is revved. As the engine warms up, fan speed (and noise) should increase noticeably. If fan noise and speed seem excessive, stop the engine and put paint marks on the fan pulley and the back of the fan clutch. Then hook up an engine timing light and start the engine. When the timing light is pointed at the back of the fan clutch, the paint marks should move relative to one another. If the paint marks stay together as engine speed is varied, the clutch is seized and must be replaced.

Front-wheel-drive vehicles are usually fitted with electrically powered fans. The fan is mounted to the back of the radiator shroud, and operates only when necessary. Some fans have both high and low speeds; others have just one speed. Fan operation is usually triggered by coolant temperature and/or A/C system operation. In some systems, a temperature sensitive fan switch is threaded into a radiator tank or an engine part to sense coolant temperature. When coolant temperature approaches the upper limit of the normal operating range, the switch contacts close to turn on the fan. When coolant temperature drops to a preset value, the switch contacts open to turn off the fan. Some switches supply power or ground directly to the fan motor; others activate a relay which then powers the fan. In other systems, fan operation is controlled by the fuel injection system computer, which obtains coolant temperature information from its engine coolant temperature sensor. When coolant temperature reaches a preset value, the computer activates the fan and continues to monitor coolant temperature. When the temperature drops to a preset value, the computer turns off the fan.

Coolant temperature switches can be normally open or normally closed, and sensor resistance specifications vary. Refer to the vehicle manufacturer's service manual to determine how the system you are working on operates.

When a vehicle is equipped with air conditioning, turning on the system usually activates the engine cooling fan automatically. In some cases, however, the fan is not turned on until refrigerant pressure reaches a preset value. Again, refer to the vehicle manufacturer's service manual for information.

Fan shrouds and air dams are an important part of a vehicle's cooling system. They should be in place and undamaged. Check the shroud fasteners to make sure that they are tight—a loose shroud could fall against a spinning fan or belts. Check the shroud for cracks, especially at the top and bottom edges. These areas can be damaged by leaning on the shroud during engine service or accidentally driving over road debris. Replace a damaged shroud. The air dam helps prevent air from flowing under the vehicle, forcing it to pass through the radiator. Check it for damage and loose fasteners. Replace a damaged air dam.

Task 11   Inspect, test, and replace auxiliary oil coolers.

Some engines, especially diesels and turbocharged engines, have an oil cooler. This oil cooler may by contained in one of the radiator tanks or mounted separately near the front of the engine or ahead of the radiator support. External oil coolers should be inspected for leaks and restricted air flow passages.

The purpose of an oil cooler is to limit maximum oil temperature. If oil temperature reaches approximately 250°F (121°C), the oil combines with oxygen in the air to form carbon and a sticky varnish. This process is known as oxidation.

An oil cooler should not be installed on vehicles that do not require one. If engine oil is "overcooled," it cannot get hot enough for unburned fuel (from blowby) and water (from condensation) to evaporate. Oil must reach a temperature of at least 215°F (102°C) for this to occur.

E. Fuel, Electrical, Ignition, and Exhaust Systems Inspection and Service (8 Questions)

Task 1   Inspect and clean or replace fuel and air induction system components, intake manifold, and gaskets.

Inspect the intake manifold for cracks, corrosion, stripped threaded holes, and other damage. Check for a plugged heat riser crossover passage and plugged EGR passages. Also check the exhaust gas crossover passage on the bottom of the manifold for cracks; this may indicate that the exhaust manifold heat riser valve is stuck shut.

When installing an intake manifold, position gaskets carefully (especially on V-type engines). An intake manifold vacuum leak may cause a cylinder misfire at idle speed, but not during acceleration when the manifold vacuum is reduced.

When installing an intake manifold that uses synthetic rubber seals at the front and rear ends, apply a dab of silicone sealer at the very ends of the seals where they meet the side gaskets. Do not apply sealer all along the top and bottom of the seals. The sealer often acts as a lubricant, allowing the seals to be squeezed out of place as the manifold bolts are tightened.

Always torque intake manifold bolts in the sequence recommended by the engine manufacturer.

Task 2   Inspect, service, or replace air filters, filter housings, and intake ductwork.

Inspect the air filter housing and any ductwork directing fresh air to the housing. Ductwork often leads from the radiator support or inner fender to the air filter housing. Make sure that this ductwork is present and undamaged.

Check the air filter housing itself. Make sure that the housing is securely mounted and hose or duct connections are tight. On carbureted and throttle body injected (TBI) engines, make sure that the gasket between the air filter housing and the carburetor or TBI unit is present and in good condition. Check the housing lid or cover to make sure that it fits properly and any seals or gasket material are in good condition. Streaks of dust or other debris around a sealing area indicate that the seal is leaking. Make sure that all clips or wing nuts securing the lid or cover are present and working properly.

Check the air filter for damage or excessive dirt. Some filters can be cleaned using compressed air; others are simply replaced. Check the filter or filter housing for an instruction label and follow the manufacturer's recommendations, if present, especially those for filter element replacement intervals.

When cleaning a pleated paper type air filter element using compressed air, hold the air gun tip about 6 inches (152 mm) from the inside or engine side of the filter and direct air against the filter material. If air is blown into the filter from the wrong direction, dirt will be lodged further into the filter material, not removed. If the air gun tip is held too close to the filter element, the element may be damaged. Wipe the inside of the filter housing with a damp cloth before installing a cleaned or new filter element.

Inspect any ductwork leading from a remote air filter to the throttle body assembly. This is especially important on fuel injected engines with remote or in-line mass air flow sensors. Leaking ductwork on these engines will allow unmeasured air into the engine, causing an incorrect air/fuel ratio. Make sure that clamps securing the ductwork to the throttle body, mass air flow meter, or air filter housing are tight.

Task 3   Inspect turbocharger/supercharger; determine necessary action.

The turbocharger/supercharger and all its mounting brackets, heat shields, and ducting should be checked for damage. Replace or repair damaged or missing components.

Check the air intake side of the turbocharger/supercharger system for leaks. If there is a leak in the intake system before the compressor housing, dirt may enter the turbocharger and damage either the compressor or turbine wheel blades. When a leak is present in the intake system between the compressor wheel housing and the cylinders, turbocharger pressure is reduced.

Turbocharger/supercharger boost pressure may be tested with a pressure gauge connected to the intake manifold. Boost pressure should be tested during hard acceleration while driving the vehicle. Excessive boost pressure is caused by a wastegate that is stuck closed, a leaking wastegate diaphragm, or a disconnected wastegate linkage. Reduced turbocharger boost pressure may be caused by a wastegate that is stuck open.

When diagnosing the cause of blue exhaust smoke on a turbocharged vehicle, first perform oil consumption diagnosis as though the engine was *not* turbocharged. While turbos are commonly blamed for excessive oil consumption problems, about half of the turbos returned under warranty are not defective. Refer to Task A.5 for more information about excessive oil consumption on turbocharged engines.

Task 4   Test battery; charge as necessary.

A battery's state of charge can be determined by measuring the specific gravity of the electrolyte solution, or "acid." To check specific gravity, remove the battery cell caps and use a hydrometer to withdraw some electrolyte from each cell. The electrolyte in a fully charged battery has a specific gravity of 1.260–1.270 when it is at a temperature of 80°F (27°C). If specific gravity is below 1.260, the battery should be charged before further testing. An important indication of battery condition is the spread or variation in specific gravity between cells. Specific gravity should not vary more than 0.050 from the lowest cell to the highest. If it does, the battery has internal damage and should be replaced.

On maintenance-free batteries, the cell caps are usually sealed to the battery case and cannot be removed. Some of these batteries have a built-in hydrometer, or "magic eye." Look down into the lens on top of the eye. If the eye appears to be green, the battery is sufficiently charged for testing. If the eye is dark, the battery should be charged before testing. If the eye is yellow or clear, the battery is low on electrolyte and should be replaced. Keep in mind that the built-in hydrometer measures specific gravity in only one cell, so a battery showing a green eye could still be defective.

Another way to determine a battery's state of charge is to measure open circuit voltage. First remove the battery surface charge by connecting a 50-ampere load across the battery terminals for 10 seconds. Then wait 10 minutes for the battery to stabilize. Disconnect both battery cables from the battery and use a voltmeter to measure voltage across the terminals. Open circuit voltage will be at least 12.6 volts on a fully charged battery *after* the surface charge has been removed. If open circuit voltage is below 12.4 volts, the battery must be charged before further testing.

The best way to determine a battery's ability to deliver power is to perform a load or capacity test. During a load test, a load tester is used to discharge the battery at one-half of its stated cold cranking ampere (CCA) rating for 15 seconds. Battery voltage is recorded at the end of this time, while the load is still applied. If the test is performed with the battery at about 70°F (21°C), a battery voltage reading of 9.6V or higher indicates that the battery is in good condition.

A discharged battery can be either fast charged or slow charged. Slow charging is always preferable if time is available to do this. Slow charging allows the chemical changes that take place during charging to occur throughout the entire thickness of the battery plates instead of on the surface of the plates only. Slow charging also lessens the chances that the battery will be overheated (and permanently damaged) during charging.

The correct slow charging rate for a battery depends on the battery's reserve capacity rating, but is usually 10 amperes or less. A battery with an 80 minute reserve capacity should be charged at 5 amperes for 10 hours or 10 amperes for 5 hours. If reserve capacity is about 150 minutes, the battery should be charges at 5 amperes for 20 hours or 10 amperes for 10 hours. Check specific gravity hourly while the battery is slow charging. When specific gravity does not increase from one check to the next, slow charging is complete.

Fast charging takes place at a much faster rate (often up to 30 amperes) and takes much less time (less than 2 hours) than slow charging. Electrolyte temperature and battery voltage should be monitored during fast charging. If electrolyte temperature reaches 125°F or voltage reaches 15V, reduce the charging rate. The correct fast charging rate depends upon battery capacity (ampere-hours) and specific gravity reading. A battery with a 55 ampere-hour capacity and a specific gravity reading of 1.200–1.225 (3/4 charged) should be fast charged for about 35 minutes. If the same battery starts out at a specific gravity of 1.125–1.150 (dead), it should be fast charged for about 80 minutes.

Task 5   Remove and replace starter.

Radio station presets should be recorded (or a memory "keep alive" tool should be connected) before beginning starter removal. Then make sure that the ignition is turned off and disconnect the battery negative cable.

Label all wires connected to the starter and disconnect the wires. Remove the starter mounting bolts and the starter. Retain any shims located between the starter motor and the cylinder block.

If the original starter motor has been repaired and is being reinstalled, install the original shims as well. If a new or rebuilt starter is being installed, temporarily mount the new part without any shims. Use a screwdriver to pry the starter motor pinion gear into the engaged position and measure clearance between the starter pinion gear and the flywheel ring gear. Check the appropriate service manual for clearance specifications, and install the necessary shims. On one manufacturer's starter, a 0.015-inch (0.38-mm) shim will increase the clearance approximately 0.005-inch (0.127-mm).

Task 6   Inspect and replace positive crankcase ventilation (PCV) system components.

Inspect the positive crankcase ventilation (PCV) system hoses for cracks, swelling, and kinking. The hose running from the carburetor base or throttle body is especially prone to damage from heat and contact with crankcase vapors.

Disconnect the hose from the carburetor or throttle body and check the hose (and carburetor or throttle body vacuum fitting) for obstructions. A restricted hose, fitting, or PCV valve may cause moisture and sludge to accumulate in the engine.

Remove the PCV valve from the rocker arm cover and the hose. Shake the valve beside your ear and listen for the pintle inside the valve to rattle. If no rattle is heard, the PCV valve should be replaced. Some manufacturers recommend removing the PCV valve from the rocker arm cover with the engine idling. If vacuum is not present at the valve inlet, the valve, hose, or vacuum fitting in the carburetor or throttle body is obstructed. Replace an obstructed hose or valve. A plugged vacuum fitting can usually be cleared using your fingers to turn the appropriate size drill bit.

Many PCV systems include a fresh air tube to supply filtered air to the crankcase. The air may be filtered by the main intake air filter or a separate PCV filter. An obstructed PCV vacuum hose or valve may allow crankcase pressure to become excessive, pushing blowby gases and vaporized oil out the fresh air tube and into the air filter. Worn piston rings can also cause this condition, since blowby is excessive.

A PCV valve can also become stuck in the open position. This allows too much air to enter the engine at idle and causes a lean air/fuel ratio. Symptoms include a rough idle and stalling. Replace a PCV valve stuck in the open position.

Task 7   Visually inspect and reinstall primary and secondary ignition system components; time distributor.

On engines fitted with a distributor, check for a cracked, worn out, or damaged distributor cap. Pull each spark plug wire from the cap (one at a time) and check for burned or corroded terminals. Check the spark plug cables for burned, pinched, cut, or oil-soaked insulation. Replace damaged cables. Remove the cap and check inside it. If the cap has excessively worn or corroded terminals, replace it. Check for carbon tracking and, if found, replace the cap. Check the high tension cable leading to the ignition coil. Ignition coils sometimes leak oil, which will soften and damage the cable. If oil is found, replace the ignition coil and the cable. Check the distributor rotor for a burned, pitted, or excessively worn contact.

On engines equipped with a point-type ignition system, inspect the points. Replace points that have excessively worn, pitted, or burned contacts, or a worn rubbing block. Inspect the distributor cam for wear and, if necessary, replace it. The mechanical advance advances spark timing as engine speed increases. Check that the advance mechanism is not seized by grasping the rotor and attempting to turn it in the direction of rotor rotation. The rotor should move in the direction of rotation against spring pressure, but not the opposite direction. When the rotor moves, pivoted weights under the rotor or breaker plate should move outward. The vacuum advance unit controls spark advance in relation to engine load. To test the advance unit, connect a hand operated vacuum pump to the hose nipple on the unit and watch the breaker plate while operating the pump. The breaker plate should rotate in the direction opposite that of the distributor rotor as the pump is operated. If the unit will not hold vacuum, the diaphragm is damaged and the vacuum advance unit must be replaced. If the unit holds vacuum, but the breaker plate does not move, the plate may be seized. Check for a rusted pivot point or a foreign object (like a dropped point retaining screw) that may be jammed between the plate and the distributor housing.

On engines equipped with solid state ignition, check the centrifugal and vacuum advance mechanisms, if equipped. Inspect the reluctor or pole piece to make sure that it is not contacting the magnetic pickup or pickup coil. Replace damaged parts.

Remove and inspect the spark plugs. Use the appropriate socket to prevent spark plug damage. Remove the plugs when the engine is cold, especially if the engine has aluminum cylinder heads. Removing plugs from an aluminum head when the engine is hot can damage the aluminum threads. If the plugs are in good condition, apply a small amount of antiseize compound to the spark plug threads. Then thread the plugs into the head by hand to avoid cross-threading. Torque the plugs to specifications.

On most, but not all, engines fitted with a distributor type ignition system, the distributor can be rotated to adjust spark timing. Vehicles produced since 1972 have an underhood emissions label that outlines the steps necessary to adjust spark timing. Follow these instructions. In a typical sequence, the engine is brought to normal operating temperature. Then the vacuum advance hose (if equipped) is disconnected and the hose is plugged. A stroboscopic timing light is connected to the #1 cylinder spark plug cable and the engine is started. Idle speed is adjusted to specifications, and then the timing light is pointed at a metal tab attached to the timing cover. If necessary, the distributor is rotated to cause a notch in the crankshaft pulley or harmonic balancer to align with a mark on the metal tab.

Task 8   Inspect, service, and replace exhaust manifold.

Exhaust manifolds can be made of cast iron or sheet metal. Sheet metal manifolds are usually made of stainless steel. Inspect exhaust manifolds for cracks and leaks. On vehicles with computer controlled fuel delivery systems, air entering the exhaust system through a crack or leak ahead of the oxygen sensor can cause driveability and emission control system problems. Replace a cracked manifold. If a leak is found at the connection to the cylinder head, remove the manifold and install a gasket. Some manifolds also use a gasket at their connection to the exhaust header pipe. Others (especially front-wheel-drive vehicles) use a ball and socket joint that allows the engine to move slightly during acceleration. Inspect this joint for leaks, too. Some engines use graphite impregnated circular seals that can be replaced if they are leaking.

The exhaust manifold on carbureted and throttle body fuel infected (TBI) engines may be equipped with a manifold heat control valve. This valve is closed when the engine is cold to direct hot exhaust gases to the underside of the intake manifold, directly under the carburetor or TBI unit. The gases heat the manifold, improving fuel vaporization in the cold engine. The valve opens as the engine warms up and the added heat is not needed. If the manifold heat control valve is stuck open or fails to close when the engine is cold, the engine may stumble during acceleration. If the valve is stuck in the closed position, engine power will be reduced and the intake manifold will overheat. The floor of the intake manifold may crack.

With the engine cold and shut off, check to see if the valve moves freely. Older engines use a bimetal thermostatic spring to operate the valve. Grasp the valve counterweight and rotate it back and forth. On newer engines the valve is opened and closed using a vacuum actuator. Connect a vacuum pump to the actuator and apply vacuum to test the valve. The valve shaft and bushings should be lubricated periodically with a special solvent that contains graphite. Check the vehicle manufacturer's service manual for lubricant information.

Sample Test

Please note the letter and number in parentheses following each question. They match the overview in section 4 that discusses the relevant subject matter. You may want to refer to the overview using this cross-referencing key to help with questions posing problems for you.

1. An engine miss is being diagnosed using a cylinder leakage test. Technician A says that any cylinder with over 20 percent leakage has excessive leakage. Technician B says that air leaking from the tailpipe indicates a cracked cylinder head. Who is right?
   A. A only
   B. B only
   C. Both A and B
   D. Neither A nor B (A.9)

2. Which of the following steps is the technician LEAST likely to perform when pressing the wrist pin into the piston and connecting rods?
   A. Align the bores in the piston and connecting rod.
   B. Heat the small end of the rod.
   C. Make sure that position marks on the piston and connecting rod are oriented properly.
   D. Heat the wrist pin. (C.12)

3. The LEAST likely reason to replace a vibration damper is if the:
   A. seal contact area on the hub is scored.
   B. inertia ring is loose.
   C. rubber is cracked.
   D. rubber is oil-soaked. (C.13)

4. While cleaning a pleated paper-type air filter element, the air gun should be held:
   A. away from the outside of the air filter element.
   B. directly against the outside of the air filter element.
   C. directly against the inside of the air filter element.
   D. away from the inside of the air filter element. (E.2)

5. If new rings are installed without removing the ring ridge, what may be the result?
   A. The piston skirt may be damaged.
   B. The piston pin may be broken.
   C. The connecting rod bearings may be damaged.
   D. The piston ring lands may be broken. (C.4)

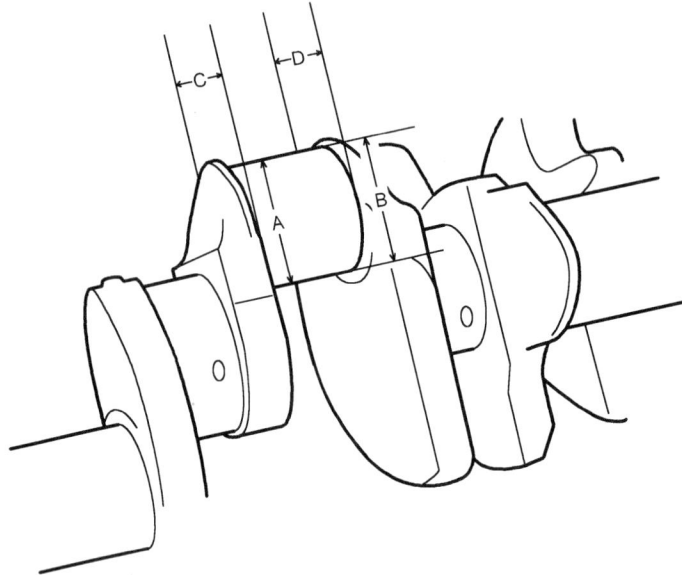

6. When measuring the crankshaft journal as shown in the figure above, the difference between measurements:
   A. A and B indicates horizontal taper.
   B. C and D indicates vertical taper.
   C. A and C indicates out-of-round.
   D. A and D indicates out-of-round.   (C.5)

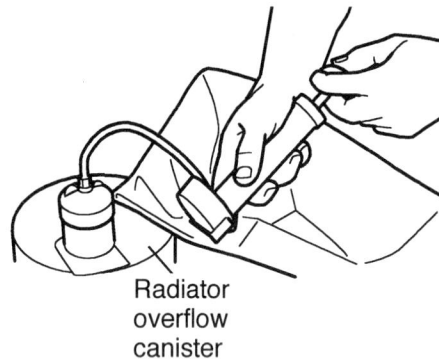

Radiator
overflow
canister

7. The tester in the figure above may be used to test all the following items EXCEPT:
   A. cooling system leaks.
   B. the radiator cap pressure relief valve.
   C. coolant specific gravity.
   D. heater core leaks.   (D.3)

8. A moderately loose alternator belt may cause:
   A. a discharged battery.
   B. a squealing noise while accelerating.
   C. poor power steering assist.
   D. engine overheating.   (D.4)

9. All of the following statements regarding manifold heat control valves are true EXCEPT:
    A. A manifold heat control valve improves fuel vaporization in the intake manifold especially when the engine is cold.
    B. A manifold heat control valve stuck in the closed position causes a loss of engine power.
    C. A manifold heat control valve stuck in the open position may cause an acceleration stumble.
    D. A manifold heat control valve stuck in the closed position reduces intake manifold temperature.   (E.8)

10. Technician A says an intake manifold vacuum leak may cause a cylinder misfire with the engine idling. Technician B says an intake manifold vacuum leak may cause a cylinder misfire during hard acceleration. Who is right?
    A. A only
    B. B only
    C. Both A and B
    D. Neither A nor B   (E.1)

11. An engine oil cooler helps to prevent:
    A. oxidation of the engine oil.
    B. excessive oil pressure.
    C. oil pump wear.
    D. main bearing wear.   (D.11)

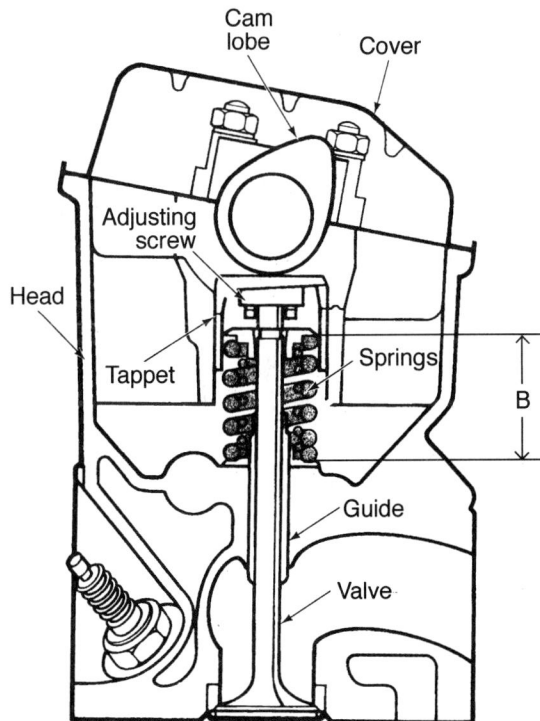

12. Measurement B in the figure above is more than specified. Technician A says this problem may bottom the lifter plunger. Technician B says a shim should be installed under the valve spring. Who is right?
    A. A only
    B. B only
    C. Both A and B
    D. Neither A nor B   (B.10)

13. Technician A says that stuck valves may cause bent pushrods. Technician B says that improper valve timing may cause bent pushrods. Who is right?
   A. A only
   B. B only
   C. Both A and B
   D. Neither A nor B   (B.11)

14. All of the following oil pump measurements should be performed with a feeler gauge EXCEPT:
   A. Measure pump cover flatness with a feeler gauge positioned between a straightedge and the cover.
   B. Measure the clearance between the inner rotor and the housing.
   C. Measure the clearance between the inner and outer rotors with the rotors installed.
   D. Measure the clearance between the top of the rotors and a straightedge positioned across the top of the oil pump.   (D.2)

15. Technician A says a warped cylinder head mounting surface on an engine block may cause valve seat distortion. Technician B says a warped cylinder head mounting surface on an engine block may cause coolant and combustion leaks. Who is right?
   A. A only
   B. B only
   C. Both A and B
   D. Neither A nor B   (C.2)

16. While discussing heli-coil installation, Technician A says the first step is to use a tap and thread the opening to match the external threads on the heli-coil. Technician B says the heli-coil should be installed with the proper size drill bit. Who is right?
   A. A only
   B. B only
   C. Both A and B
   D. Neither A nor B   (C.3)

17. When measuring valve stem-to-guide clearance, Technician A says the valve stems and guides should be measured at three vertical locations. Technician B says the valve guide diameter should be measured with either a hole or snap gauge. Who is right?
   A. A only
   B. B only
   C. Both A and B
   D. Neither A nor B   (B.6)

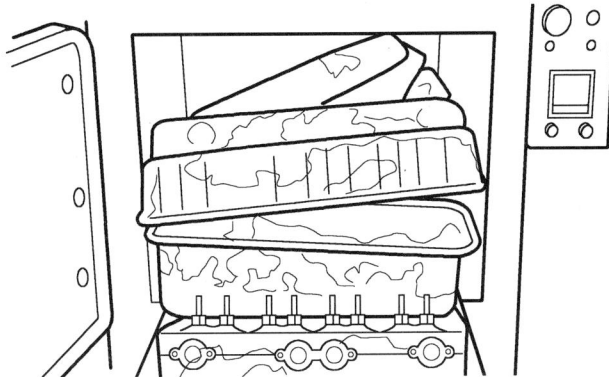

18. In the figure above, a thermal cleaner is being used to prepare iron and steel components for inspection. The thermal cleaner heats the parts to:
    A. 200 to 300°F (93 to 149°C)
    B. 550 to 700°F (288 to 371°C)
    C. 700 to 900°F (371 to 482°C)
    D. 650 to 800°F (343 to 427°C)   (C.1)

19. All of the following are causes of low engine oil pressure EXCEPT:
    A. worn camshaft bearings.
    B. worn crankshaft bearings.
    C. weak oil pressure regulator spring tension.
    D. restricted pushrod oil passages.   (D.1)

20. Technician A says metal burrs on the crankshaft flange may cause excessive wear on the ring gear and starter drive gear teeth. Technician B says metal burrs on the crankshaft flange may cause improper torque converter-to-transmission alignment. Who is right?
    A. A only
    B. B only
    C. Both A and B
    D. Neither A nor B   (C.14)

21. A bent connecting rod may cause:
    A. uneven connecting rod bearing wear.
    B. uneven main bearing wear.
    C. uneven piston pin wear.
    D. excessive cam bearing wear.   (C.11)

22. Technician A says that the timing cover must be removed in order to replace the crankshaft front oil seal. Technician B says that the lip on the front oil seal must face toward the crankshaft pulley hub or harmonic balancer. Who is right?
    A. A only
    B. B only
    C. Both A and B
    D. Neither A nor B   (C.15)

23. During a compression test, a cylinder has 40 percent of the specified compression reading. When the technician performs a wet test, the compression reading on this cylinder has 75 percent of the specified reading. The cause of the low compression reading could be:

    A. worn piston rings.

    B. a burned exhaust valve.

    C. a bent intake valve.

    D. a worn camshaft lobe.   (A.8)

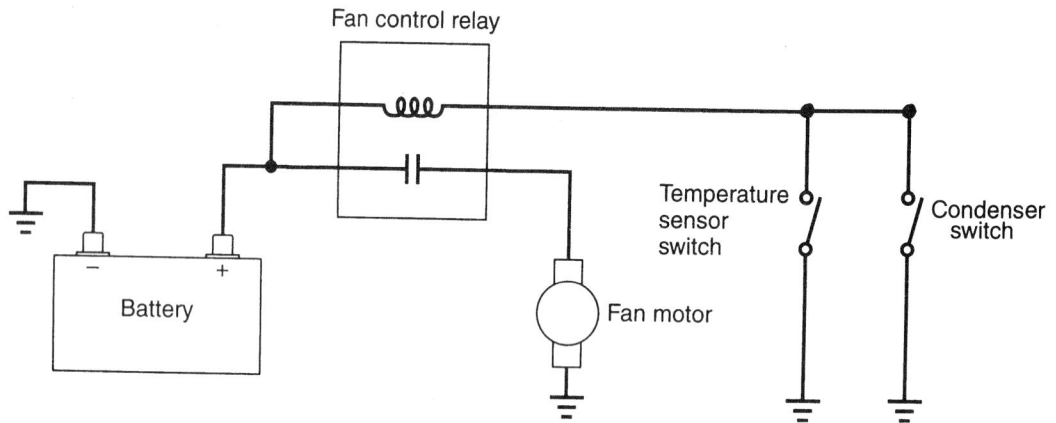

24. In the figure above, an open ground circuit on the engine temperature sensor switch may cause:

    A. continual cooling fan motor operation.

    B. a completely inoperative cooling fan motor.

    C. a burned-out cooling fan motor.

    D. engine overheating.   (D.10)

25. The LEAST likely place an RTV sealer would be used is the:

    A. oil pan.

    B. valve cover.

    C. intake manifold.

    D. cylinder head.   (C.16)

26. A collapsed cooling system hose may be an indication of a:

    A. damaged radiator cap sealing gasket.

    B. damaged radiator filler neck seat.

    C. damaged expansion tank.

    D. damaged radiator cap vacuum valve.   (D.9)

27. While discussing cooling system service, Technician A says if the cooling system pressure is reduced, the coolant boiling point is increased. Technician B says when more antifreeze is added to the coolant (up to an 80/20 mix), the coolant boiling point is increased. Who is right?

    A. A only

    B. B only

    C. Both A and B

    D. Neither A nor B   (D.7)

28. Technician A says a defective water pump bearing may cause a growling noise when the engine is idling. Technician B says the water pump bearing may be ruined by coolant leaking past the pump seal. Who is right?
    A. A only
    B. B only
    C. Both A and B
    D. Neither A nor B   (D.8)

29. Technician A says worn valve stem seals may cause rapid valve stem and guide wear. Technician B says worn valve stem seals may cause excessive oil consumption. Who is right?
    A. A only
    B. B only
    C. Both A and B
    D. Neither A nor B   (B.5)

30. While measuring valve springs, Technician A says the valve spring must be rotated while measuring squareness. Technician B says that spring squareness can be checked by rolling the spring on a surface plate. Who is right?
    A. A only
    B. B only
    C. Both A and B
    D. Neither A nor B   (B.3)

31. Technician A says worn valve lock grooves may cause the valve locks to fly out of place with the engine running, resulting in severe engine damage. Technician B says worn valve lock grooves may cause a clicking noise with the engine idling. Who is right?
    A. A only
    B. B only
    C. Both A and B
    D. Neither A nor B   (B.4)

32. While adjusting mechanical valve lifters, Technician A says when the valve clearance is checked on a cylinder, the piston in that cylinder should be at top dead center (TDC) on the exhaust stroke. Technician B says some mechanical valve lifters have removable shim pads available in various thicknesses to provide the proper valve clearance. Who is right?
    A. A only
    B. B only
    C. Both A and B
    D. Neither A nor B   (B.13)

33. When measuring main bearing bores, provided the vertical reading is not the largest, out-of-round measurements are acceptable if less than:
    A. 0.001 inch (0.025 mm).
    B. 0.010 inch (0.254 mm).
    C. 0.005 inch (0.127 mm).
    D. 0.015 inch (0.381 mm).   (C.6)

34. When measuring crankshaft end play, it is LEAST likely that a technician would use:
    A. a micrometer.
    B. a pry bar.
    C. a dial indicator.
    D. feeler gauges.   (C.7)

35. The customer complains that the engine cranks but does not start; the first thing to check should be:
    A. valve train operation.
    B. battery voltage.
    C. compression.
    D. engine vacuum.   (A.2)

36. Technician A says that many thermostats are marked to indicate which way coolant should flow through them. Technician B says a thermostat that is not marked to indicate coolant flow direction should be installed so that the thermal element is pointed toward the radiator. Who is right?
    A. A only
    B. B only
    C. Both A and B
    D. Neither A nor B   (D.6)

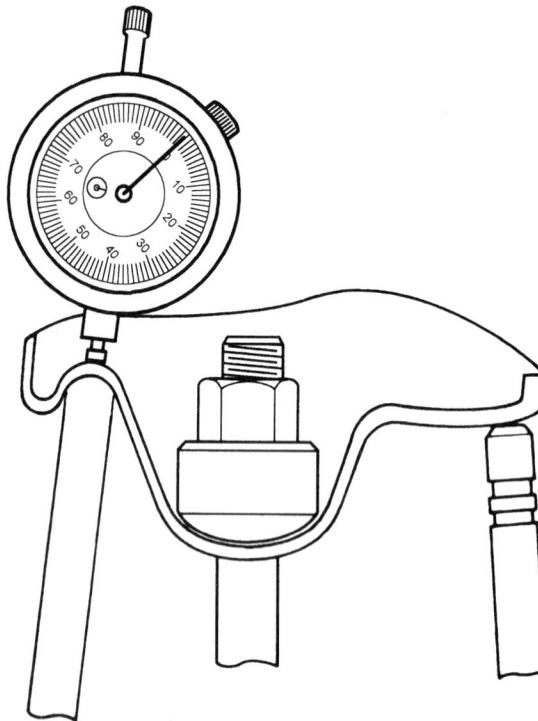

37. The tool in the figure above is being used to check the:
    A. valve lifter height.
    B. camshaft journal out-of-round.
    C. pushrod length.
    D. camshaft lobe lift.   (B.15)

38. When discussing camshaft bearing clearance, Technician A says excessive camshaft bearing clearance may result in lower-than-specified oil pressure. Technician B says excessive camshaft bearing clearance may cause a clicking noise when the engine is idling. Who is right?
    A. A only
    B. B only
    C. Both A and B
    D. Neither A nor B   (B.16)

39. On engines where the camshaft drive gear teeth mesh directly with the crankshaft gear teeth, Technician A says the timing gear backlash may be measured with a dial indicator. Technician B says on this type of engine, the timing gear backlash may be measured with a micrometer. Who is right?
    A. A only
    B. B only
    C. Both A and B
    D. Neither A nor B   (B.14)

40. During a cylinder balance test on an engine with electronic fuel injection, one cylinder provides very little rpm drop. Technician A says the ignition system may be misfiring on that cylinder. Technician B says the engine may have an intake manifold vacuum leak. Who is right?
    A. A only
    B. B only
    C. Both A and B
    D. Neither A nor B   (A.7)

41. Technician A says improper valve timing may cause reduced engine power. Technician B says improper valve timing may cause bent valves in some engines. Who is right?
    A. A only
    B. B only
    C. Both A and B
    D. Neither A nor B   (B.17)

42. Technician A says hydraulic valve lifter bottoms should be flat or concave. Technician B says a sticking lifter plunger may cause a burned exhaust valve. Who is right?
    A. A only
    B. B only
    C. Both A and B
    D. Neither A nor B   (B.12)

43. When removing the timing belt from an overhead cam engine, the technician must first:
    A. remove the rocker arm assembly.
    B. mark the timing belt for position.
    C. mark the timing belt for direction.
    D. remove the water pump.   (B.1)

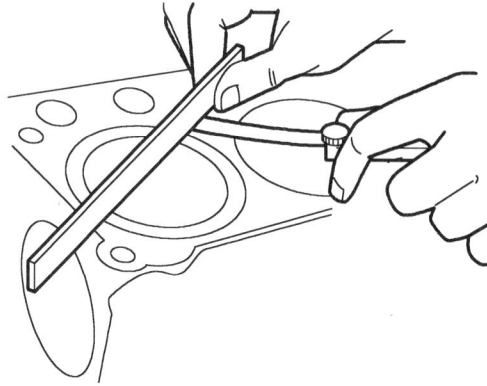

44. An OHC cylinder head is being inspected and the feeler gauge shown in the figure above is 0.014 inch (0.36 mm) thick. Technician A says the head should be resurfaced and reinstalled. Technician B says that warpage on the cam side must be measured to determine if the head is usable. Who is right?
    A. A only
    B. B only
    C. Both A and B
    D. Neither A nor B   (B.2)

45. The hose from the positive crankcase ventilation (PCV) valve to the intake manifold is restricted. This problem could result in:
    A. an acceleration stumble.
    B. oil accumulation in the air cleaner.
    C. engine surging at high speed.
    D. engine detonation during acceleration.   (E.6)

46. In the figure above, the technician is most likely checking:
    A. valve guide depth.
    B. valve seat angle.
    C. cylinder head flatness.
    D. valve seat runout.   (B.9)

47. Technician A says that a soft or gummy heater hose may be caused by a missing exhaust manifold heat shield. Technician B says that a brittle or hard lower radiator hose may be caused by engine oil leaking onto the hose. Who is right?
    A. A only
    B. B only
    C. Both A and B
    D. Neither A nor B   (D.5)

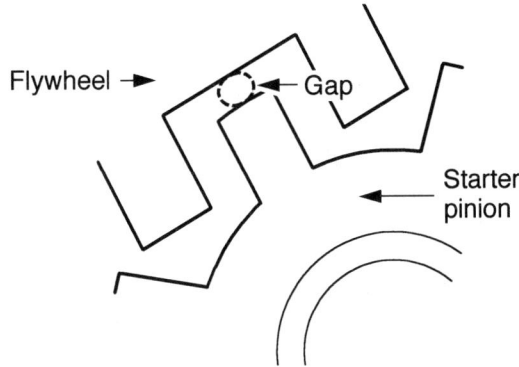

48. When installing a starter motor, shims may be required by some manufacturers to adjust the gap between the starter pinion gear and flywheel ring gear. Adding a 0.015 inch (0.381 mm) shim between the starter motor and the engine block will:
    A. decrease the clearance by 0.005 inch (0.127 mm).
    B. decrease the clearance by 0.015 inch (0.381 mm).
    C. increase the clearance by 0.010 inch (0.254 mm).
    D. increase the clearance by 0.005 inch (0.127 mm).   (E.5)

49. Technician A says that when inspecting piston ring grooves for wear, you should place a new ring in the groove and then insert a feeler gauge between the ring and the groove. Technician B says that you should simply insert a feeler gauge in the ring groove and then compare the measurement to specifications. Who is right?
    A. A only
    B. B only
    C. Both A and B
    D. Neither A nor B   (C.10)

50. Reduced turbocharger boost pressure may be caused by a:
    A. wastegate valve stuck closed.
    B. wastegate valve stuck open.
    C. leaking wastegate diaphragm.
    D. disconnected wastegate linkage.   (E.3)

51. While discussing torque-to-yield head bolts, Technician A says compared to conventional head bolts, torque-to-yield bolts provide a more uniform clamping force. Technician B says torque-to-yield bolts are tightened to a specific torque and then rotated tighter a certain number of degrees. Who is right?
    A. A only
    B. B only
    C. Both A and B
    D. Neither A nor B   (B.18)

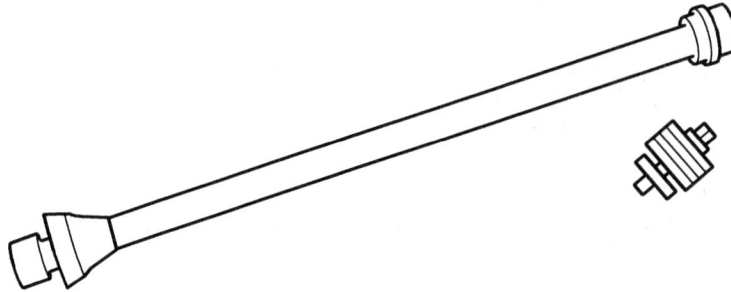

52. The tool shown in the figure above is used to:
    A. remove camshaft bearings only.
    B. install camshaft bearings only.
    C. measure camshaft bearing alignment.
    D. both remove and install camshaft bearings.   (C.8)

53. Technician A says improper balance shaft timing causes severe engine vibrations. Technician B says the balance shafts are timed in relation to the camshaft. Who is right?
    A. A only
    B. B only
    C. Both A and B
    D. Neither A nor B   (C.9)

54. While discussing basic diagnostic procedures, Technician A says the most complicated diagnostic tests should be performed first. Technician B says that you should first question the customer to obtain as much information as possible about the problem. Who is right?
    A. A only
    B. B only
    C. Both A and B
    D. Neither A nor B   (A.1)

55. The following statements about distributor advances are true EXCEPT:
    A. The vacuum advance controls spark advance in relation to engine load.
    B. The mechanical advance controls spark advance in relation to engine speed.
    C. The mechanical advance rotates the reluctor in the opposite direction of the shaft rotation.
    D. The vacuum advance rotates the pickup plate in the opposite direction to shaft rotation.   (E.7)

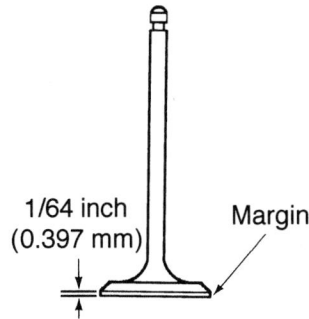

56. A valve margin of 1/64 inch (0.397 mm) as shown above may cause:
    A. a clicking noise at idle speed.
    B. valve overheating and burning.
    C. improper valve seating.
    D. valve seat recession.   (B.7)

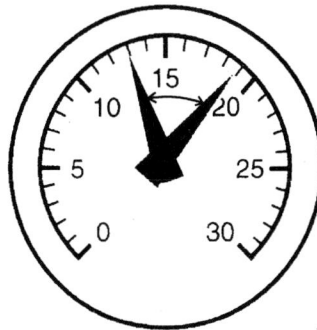

57. With the engine idling, a vacuum gauge connected to the intake manifold fluctu-
    ates as shown in the figure above. These vacuum gauge fluctuations may be
    caused by:
    A. late ignition timing.
    B. intake manifold vacuum leaks.
    C. a restricted exhaust system.
    D. sticky valve stems and guides.   (A.6)

58. A battery rated at 600 cold cranking amps (cca) is load tested at one-half of its
    rated cca for 15 seconds. The results show 10.1 volts. The results indicate that this
    battery:
    A. is satisfactory.
    B. needs recharging.
    C. is bad and should be replaced.
    D. should be retested at load for 30 seconds.   (E.4)

59. Oil is leaking from the crankshaft rear main bearing seal on an engine. Technician A says the oil seal could be faulty. Technician B says the PCV system may not be functioning. Who is right?
    A. A only
    B. B only
    C. Both A and B
    D. Neither A nor B   (A.3)

60. A high-pitched squealing noise is heard during hard acceleration. This noise may be caused by:
    A. an intake manifold leak.
    B. the choke stuck closed.
    C. a fuel system leak.
    D. a small leak in the exhaust manifold.   (A.5)

61. A heavy thumping noise occurs with the engine idling, but the oil pressure is normal. This noise may be caused by:
    A. worn pistons and cylinders.
    B. loose flywheel bolts.
    C. worn main bearings.
    D. loose camshaft bearings.   (A.4)

62. Technician A says removable valve seat inserts may be removed with a special puller or a pry bar. Technician B says a special driver is used to install the valve seat insert, and the insert should be staked after installation. Who is right?
    A. A only
    B. B only
    C. Both A and B
    D. Neither A nor B   (B.8)

6 Additional Test Questions for Practice

Additional Test Questions

Please note the letter and number in parentheses following each question. They match the overview in section 4 that discusses the relevant subject matter. You may want to refer to the overview using this cross-referencing key to help with questions posing problems for you.

1. Technician A says valve guide height is measured from the bottom of the valve guide to the top of the valve guide. Technician B says increased oil consumption may result from excessive valve stem-to-guide clearance. Who is right?
   A. A only
   B. B only
   C. Both A and B
   D. Neither A nor B   (B.6)

2. The LEAST likely cause of excessive blue smoke in the exhaust of a turbocharged engine is:
   A. a PCV valve stuck in the open position.
   B. worn turbocharger seals.
   C. worn valve guide seals.
   D. worn piston rings.   (E.3)

3. An engine is being disassembled after spinning a rod bearing. Technician A says to flush the oil passages in the head and block. Technician B says to replace the oil cooler. Who is right?
   A. A only
   B. B only
   C. Both A and B
   D. Neither A nor B   (C.1)

4. A bent valve spring must be replaced if it has a spring height variance of more than:
   A. 0.005 inch (0.127 mm).
   B. 0.0625 inch (1.588 mm).
   C. 0.125 inch (3.175 mm).
   D. 0.025 inch (0.635 mm).   (B.3)

5. All of the following could cause a bent pushrod EXCEPT:
   A. worn cam bearings.
   B. a broken timing chain.
   C. a sticking valve.
   D. improper valve adjustment.   (B.11)

6. Technician A says that room temperature vulcanizing (RTV) sealant is used to secure threaded fasteners. Technician B says that fumes from an anaerobic sealant can damage an oxygen sensor. Who is right?
   A. A only
   B. B only
   C. Both A and B
   D. Neither A nor B   (C.16)

7. On an overhead cam (OHC) engine with removable bearing caps, which of the following is used to measure bearing alignment?
   A. A straightedge
   B. Plastigauge
   C. A dial indicator
   D. A telescoping gauge   (B.16)

8. Technician A says that balance shafts should be checked for runout following the same procedure used for measuring camshaft runout. Technician B says that balance shaft journals should be measured for taper following the same procedure used for measuring crankshaft journal taper. Who is right?
   A. A only
   B. B only
   C. Both A and B
   D. Neither A nor B   (C.9)

9. Technician A says that a noisy water pump could be caused by a corroded bearing. Technician B says that this could be caused by a defective seal. Who is right?
   A. A only
   B. B only
   C. Both A and B
   D. Neither A nor B   (D.8)

10. An electric cooling fan is inoperative. Technician A says this could be caused by a bad ground in the cooling fan circuit. Technician B says this could be caused by a bad wire to the fan relay. Who is right?
    A. A only
    B. B only
    C. Both A and B
    D. Neither A nor B   (D.10)

11. Technician A says that the mechanical advance weights in a distributor cause the pickup or breaker plate to rotate in a direction opposite that of distributor shaft rotation. Technician B says that the vacuum advance rotates the distributor cam or reluctor in a direction opposite that of distributor shaft rotation. Who is right?
    A. A only
    B. B only
    C. Both A and B
    D. Neither A nor B   (E.7)

12. A technician is servicing an overhead camshaft (OHC) engine where the camshaft runs without bearings in its bore. If camshaft journal-to-bore clearance exceeds specification, the technician must:
    A. replace bearings with oversized bearings.
    B. replace the camshaft.
    C. insert bushings.
    D. replace the cylinder head.   (C.8)

13. A new starter motor has been installed on an engine and the technician installed the original shim between the starter and the block. When the engine is cranked, a loud whining noise is heard. Technician A says that there is too little clearance between the starter pinion and the flywheel. Technician B says that there is too much clearance. Who is right?
    A. A only
    B. B only
    C. Both A and B
    D. Neither A nor B   (E.5)

14. With the engine at normal operating temperature, the oil pressure test is usually performed at idle speed and a higher speed such as:
    A. 1,500 rpm.
    B. 2,000 rpm.
    C. 2,500 rpm.
    D. 3,000 rpm.   (D.1)

15. Technician A says that the spark timing adjustment procedures for a vehicle can be found on an underhood label for all cars built since 1972. Technician B says that on any engine with a distributor, spark timing can be adjusted by rotating the distributor with respect to the cylinder block (or cylinder head). Who is right?
    A. A only
    B. B only
    C. Both A and B
    D. Neither A nor B   (E.7)

16. Technician A says piston ring grooves should be cleaned by using a file. Technician B says to position a feeler gauge between each ring and the ring groove to measure the ring groove clearance. Who is right?
    A. A only
    B. B only
    C. Both A and B
    D. Neither A nor B   (C.10)

17. A starter is being installed in a vehicle. Technician A says that all starters require shims to be installed. Technician B says measurements must be taken to determine the need for shims on certain vehicles only. Who is right?
    A. A only
    B. B only
    C. Both A and B
    D. Neither A nor B   (E.5)

18. Technician A says that replacing a 180°F (82°C) thermostat with a 195°F (91°C) thermostat will cause the engine to warm up faster. Technician B says that removing the thermostat from an engine may cause "hot spots" to develop in the engine. Who is right?
    A. A only
    B. B only
    C. Both A and B
    D. Neither A nor B   (D.6)

19. If either the radiator pressure cap sealing gasket or the radiator filler neck seat are damaged, which of the following is LEAST likely to occur?
    A. The lower radiator hose will burst.
    B. The engine coolant will boil.
    C. The engine coolant will overflow.
    D. The engine will overheat.   (D.9)

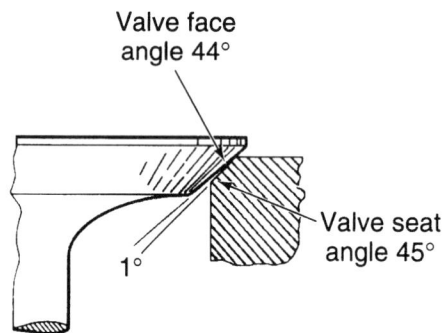

Valve face angle 44°
Valve seat angle 45°
1°

20. In the above figure, Technician A says that a three-angle valve job is shown. Technician B says that poor valve face to valve seat orientation is shown. Who is right?
    A. A only
    B. B only
    C. Both A and B
    D. Neither A nor B   (B.8)

21. After a vehicle is parked overnight and then started in the morning, the engine has a lifter noise that disappears after running for a short while. The most likely cause would be:
    A. low oil pressure.
    B. low oil level.
    C. a worn lifter bottom.
    D. excessive lifter leak-down.   (A.4)

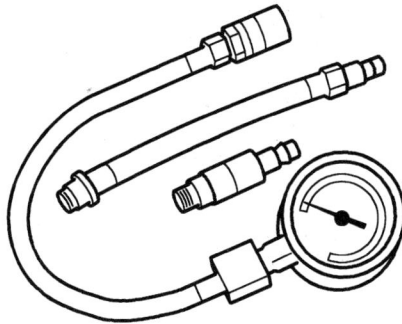

22. When using a compression tester, as shown above, the compression readings on the cylinders are all even, but lower than the specified compression. This could indicate:
    A. a blown head gasket.
    B. carbon buildup.
    C. a cracked head.
    D. worn rings and cylinders.   (A.8)

23. Technician A says that an overtensioned V-belt can damage the alternator front bearing. Technician B says that an overtensioned V-belt can cause the upper half of the crankshaft front main bearing to wear prematurely. Who is right?
    A. A only
    B. B only
    C. Both A and B
    D. Neither A nor B   (D.4)

24. Air filters should be replaced:
    A. after every 12,000 miles.
    B. after every 15,000 miles.
    C. after every 3,000 miles.
    D. according to manufacturers' recommendations.   (E.2)

25. Valve spring installed height is measured between the lower edge of the top retainer and the:
    A. cylinder head.
    B. top edge of the top shim.
    C. bottom edge of the bottom shim.
    D. spring seat.   (B.10)

26. A technician is preparing to install new oil gallery plugs in a cylinder block. Which of these operations is he LEAST likely to perform?
    A. Run a bottoming tap into the threaded gallery holes.
    B. Apply oil resistant sealer to the new plugs.
    C. Apply teflon tape to the threaded plugs.
    D. Run a rifle brush through the galleries.   (C.3)

27. The technician is using a torque wrench operated valve spring tester to measure valve spring tension. The technician pulls on the torque wrench until a click or ping is heard. The reading at this point then needs to be multiplied by:
    A. two.
    B. three.
    C. four.
    D. five.   (B.3)

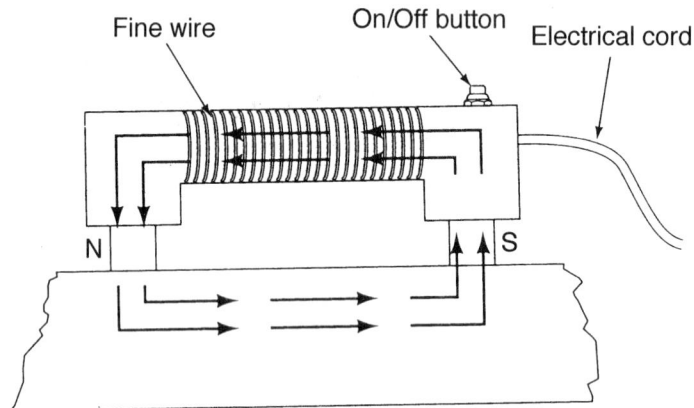

28. An electromagnetic-type tester, as shown in the figure above, and iron fillings may be used to check for cracks in:
    A. aluminum cylinder heads.
    B. pistons.
    C. cast-iron cylinder heads.
    D. aluminum intake manifolds.   (B.2)

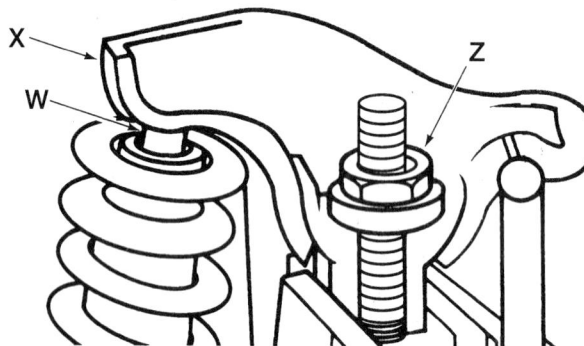

29. In the above figure, how is valve lash adjusted?
    A. By adding shims to point W
    B. By adding shims to point X
    C. No adjustment is required
    D. By turning nut Z   (B.13)

30. Which of the following would LEAST likely require crankshaft grinding?
    A. Excessive taper
    B. An out-of-round journal
    C. Excessive journal scoring
    D. Excessive thrust wear   (C.5)

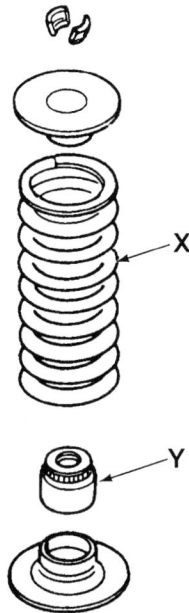

31. Technician A says that in the above figure, X can be replaced without removing the head. Technician B says Y can be replaced without removing the head. Who is right?
    A. A only
    B. B only
    C. Both A and B
    D. Neither A nor B   (B.1)

32. The following specific gravity readings were taken from a battery at 80°F (27°C): Cell 1: 1.200; Cell 2: 1.210; Cell 3: 1.190; Cell 4: 1.200; Cell 5: 1.205; and Cell 6: 1.200. What action should be taken?
    A. Load test the battery.
    B. Replace the battery.
    C. Refill the battery with fresh electrolyte.
    D. Recharge the battery.   (E.4)

33. Technician A says that when installing cam bearings in the cylinder block, each bearing must be driven into its bore until the front edge of the bearing is flush with the front edge of the bore. Technician B says that the rear edge of each bearing must be flush with the rear edge of the bore. Who is right?
    A. A only
    B. B only
    C. Both A and B
    D. Neither A nor B   (C.8)

34. Technician A says that balance shafts always rotate at crankshaft speed. Technician B says that balance shafts are found only on four-cylinder engines. Who is right?
    A. A only
    B. B only
    C. Both A and B
    D. Neither A nor B   (C.9)

35. Technician A says thermal cleaners heat parts from 1,000 to 1,500°F (538 to 816°C) to oxidize the contaminants. Technician B says aluminum parts can be cleaned in thermal cleaners without damage. Who is right?
    A. A only
    B. B only
    C. Both A and B
    D. Neither A nor B   (C.1)

36. A cylinder balance test is being performed on a engine to determine which cylinder is causing a "miss." Technician A says that when the faulty cylinder is disabled, engine rpm will drop more than for the other cylinders. Technician B says disabling the faulty cylinder will cause the engine to stall. Who is right?
    A. A only
    B. B only
    C. Both A and B
    D. Neither A nor B   (A.7)

37. All lifters in an engine are cupped (concave). Technician A says replace the camshaft. Technician B says replace the lifters. Who is right?
    A. A only
    B. B only
    C. Both A and B
    D. Neither A nor B   (B.12)

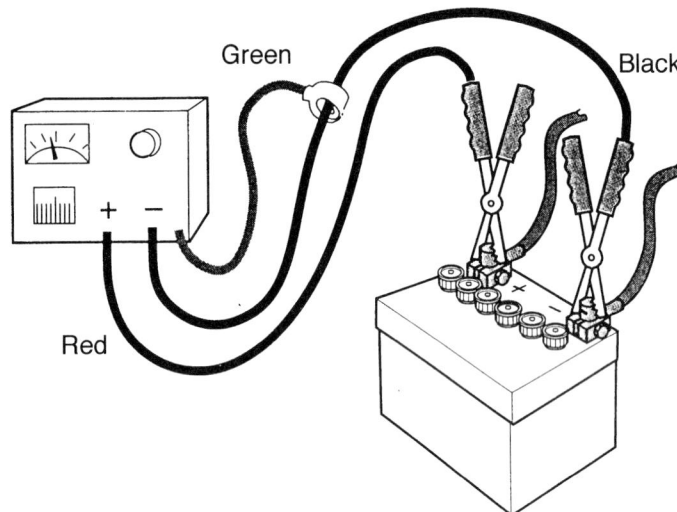

38. A battery load or capacity tester, as shown in the figure above, is used to test the battery capacity. During this test, the battery is discharged at one-half the cold cranking amperes for:
    A. 15 seconds.
    B. 20 seconds.
    C. 25 seconds.
    D. 30 seconds.   (E.4)

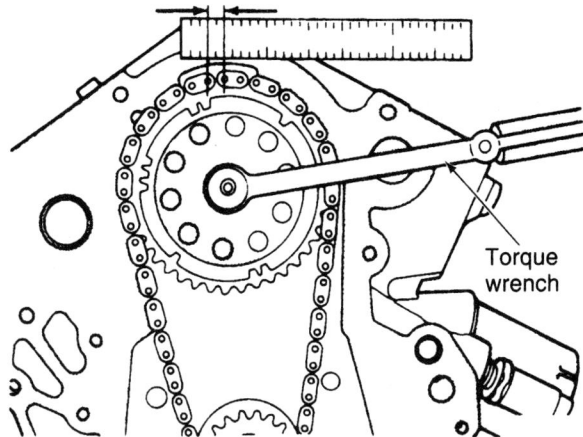

39. In the above figure, what is being performed?
    A. Adjusting cam timing
    B. Locating TDC
    C. Measuring timing chain stretch
    D. Adjusting valve lash   (B.14)

40. A cylinder block has just been hot tanked and is ready for inspection. Technician A says that the block deck should be checked for warpage using a straightedge and a feeler gauge. Technician B says that minor nicks or burrs on the block deck can be removed using a whetstone or a file. Who is right?
    A. A only
    B. B only
    C. Both A and B
    D. Neither A nor B   (C.2)

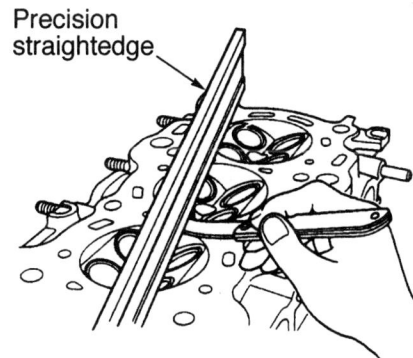

41. As shown in the above figure, cylinder head warpage is being checked on a four-cylinder engine. The head is not excessively warped if the feeler gauge being inserted is up to:
    A. 0.004 inch (0.102 mm) thick.
    B. 0.006 inch (0.152 mm) thick.
    C. 0.008 inch (0.203 mm) thick.
    D. 0.012 inch (0.305 mm) thick.   (B.2)

42. After performing a compression test on a V-8 engine, two cylinders have pressure readings of 60 psi (44 kPa) while the others have a reading of 135 psi (931 kPa). The two low cylinders are next to each other. Technician A says this could be caused by a loose timing chain. Technician B says a leaking head gasket could cause this. Who is right?
   A. A only
   B. B only
   C. Both A and B
   D. Neither A nor B   (A.8)

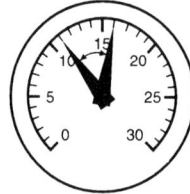

43. An engine is idling at 750 rpm. The pointer on the vacuum gauge in the figure above is floating between 11 and 16 in. Hg. The most likely cause would be:
   A. retarded timing.
   B. advanced timing.
   C. a stuck EGR valve.
   D. too lean an idle mixture.   (A.6)

44. A defective water pump can be diagnosed by all of the following EXCEPT by:
   A. observing residue at the water pump drain hole.
   B. observing a coolant leak from the water pump.
   C. hearing a groaning noise at cruising speeds.
   D. using a pressure tester.   (D.8)

45. When deciding on which sealer to use, Technician A says room temperature vulcanizing (RTV) sealer dries in the presence of air by expelling moisture into the air. Technician B says anaerobic sealer dries in the absence of air. Who is right?
   A. A only
   B. B only
   C. Both A and B
   D. Neither A nor B   (C.16)

Fan control relay

Temperature sensor switch

Condenser switch

Battery

Fan motor

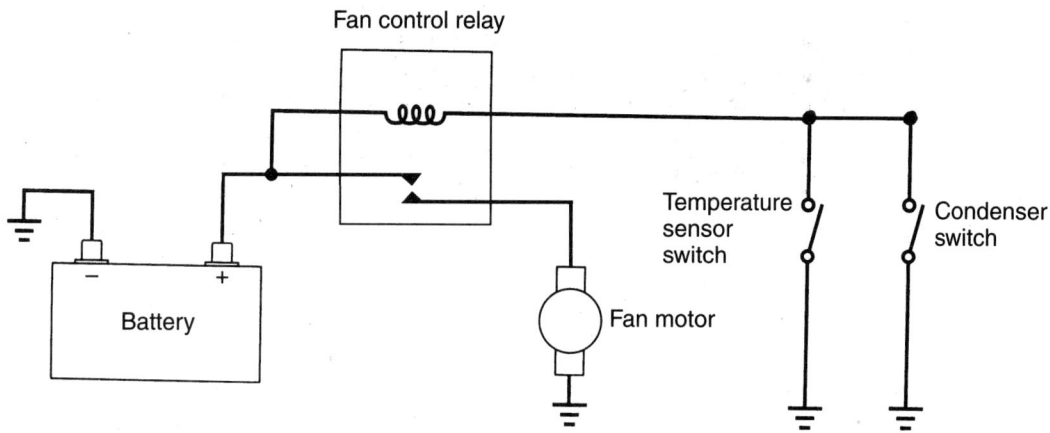

46. An electric drive cooling fan circuit is shown above. Technician A says if the coolant temperature sensor switch is stuck closed, the cooling fan will stop when the ignition is turned off. Technician B says when the air conditioner (A/C) is turned on, the fan relay winding will be grounded through the condenser switch to activate the cooling fan. Who is right?
   A. A only
   B. B only
   C. Both A and B
   D. Neither A nor B   (D.10)

47. Technician A says that some head bolts stretch permanently when they are tightened. Technician B says some head bolts cannot be reused. Who is right?
   A. A only
   B. B only
   C. Both A and B
   D. Neither A nor B   (B.18)

48. Technician A says that if a vehicle uses unleaded gasoline, an interference angle must be used on the valves. Technician B says an interference angle creates a poor seal when the engine is first started after reconditioning valves. Who is right?
   A. A only
   B. B only
   C. Both A and B
   D. Neither A nor B   (B.7)

49. When measuring main bearing bores, Technician A says the vertical measurement should not be larger than any of the others. Technician B says out-of-round measurements less than 0.001 inch (0.025 mm) are acceptable if the vertical reading is not the largest. Who is right?
    A. A only
    B. B only
    C. Both A and B
    D. Neither A nor B   (C.6)

50. Technician A says that stretched main bearing bores can be corrected by filing the main bearing caps. Technician B says this problem can be corrected by replacing the main bearing caps. Who is right?
    A. A only
    B. B only
    C. Both A and B
    D. Neither A nor B   (C.6)

51. When checking connecting rods for damage and wear, which of the following is LEAST likely to be checked?
    A. Rod center-to-center length
    B. Rod straightness
    C. Small end bore condition
    D. Big end bore out-of-round   (C.11)

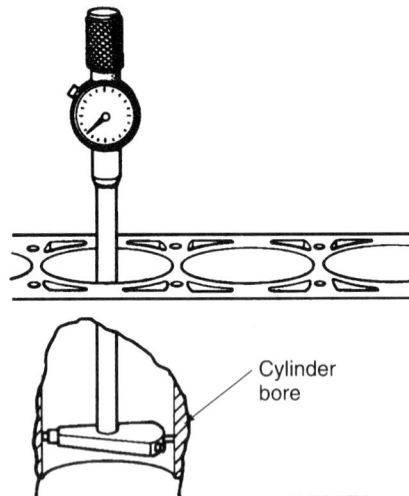

Cylinder bore

52. The gauge in the figure above is being used to check cylinder diameter near the bottom of the cylinder. If the technician wants to determine cylinder taper, where must he take an additional measurement?
    A. At the top of the cylinder, just above the ring ridge
    B. At the top of the cylinder, just below the ring ridge
    C. Near the top of the cylinder, at the top of the oil ring contact area
    D. Near the bottom of the cylinder, at 90° to the first measurement   (C.4)

53. A cylinder block deck is being measured for warpage. Technician A says that if the manufacturer does not provide a warpage limit specification, warpage under 0.005 inch (0.127 mm) is acceptable. Technician B says that if warpage exceeds 0.001 inch (0.025 mm) the block must be resurfaced.
    A. A only
    B. B only
    C. Both A and B
    D. Neither A nor B   (C.2)

54. Technician A says that some engines are fitted with an oil pump that slides over the crankshaft snout and bolts to the front of the block. Technician B says that some engines are fitted with a pump that is driven by the timing belt. Who is right?
    A. A only
    B. B only
    C. Both A and B
    D. Neither A nor B   (D.2)

55. Technician A says that compression rings should never be installed by "spiraling" them onto the piston. Technician B says that the top compression ring should always be installed on the piston first. Who is right?
    A. A only
    B. B only
    C. Both A and B
    D. Neither A nor B   (C.12)

56. The rocker arms on a pushrod engine have a 1.5:1 ratio. This means that:
    A. a cam lift of 0.250 inch (6.35 mm) will cause the valve to open 0.188 inch (4.76 mm).
    B. a cam lift of 0.250 inch (6.35 mm) will cause the valve to open 0.375 inch (9.53 mm).
    C. the engine must be fitted with hydraulic lifters.
    D. the engine must be fitted with roller lifters.   (B.11)

57. Valve face and seat concentricity can be measured with all of the following EXCEPT:
    A. a concentricity tester.
    B. a dial caliper.
    C. blue dye.
    D. a dial indicator.   (B.9)

58. The areas around the mounting holes on a sheet metal rocker cover are dished from having the fasteners overtightened. To prevent oil leaks from occurring when the rocker cover is installed, the technician should:
    A. replace the rocker cover with a new one.
    B. hammer the dished areas flat again.
    C. use two gaskets instead of one.
    D. use RTV sealant instead of a gasket.   (C.15)

59. Technician A says that valve lock grooves on the valve stems must be inspected for rounded shoulders. Technician B says valve stems having rounded or uneven shoulders require machining. Who is right?
    A. A only
    B. B only
    C. Both A and B
    D. Neither A nor B   (B.4)

60. Most manufacturers recommend that piston diameter be measured at 90° to the wrist pin bore:
    A. at the very top of the piston.
    B. at the wrist pin bore centerline.
    C. about 3/4 inch below the wrist pin bore centerline.
    D. about 1/4 inch from the bottom of the piston skirt.   (C.10)

61. A crankshaft journal is being measured in the same direction at opposite ends. This is measuring for:
    A. taper.
    B. out-of-roundness.
    C. wear.
    D. bearing size.   (C.5)

62. Technician A says that flywheel runout is checked using a dial indicator while the flywheel is still mounted on the crankshaft. Technician B says that flywheel runout is checked using a dial indicator and stand after removing the flywheel from the engine and placing it on a surface plate. Who is right?
    A. A only
    B. B only
    C. Both A and B
    D. Neither A nor B   (C.14)

63. The first step a technician should take toward determining the cause of a problem is:
    A. think of possible causes of the problem.
    B. question the customer.
    C. road test the vehicle.
    D. listen to the customer.   (A.1)

64. Vibration damper rubber should be inspected for all of the following EXCEPT:
    A. hub contact area scoring.
    B. looseness.
    C. cracks.
    D. oil soaking.   (C.13)

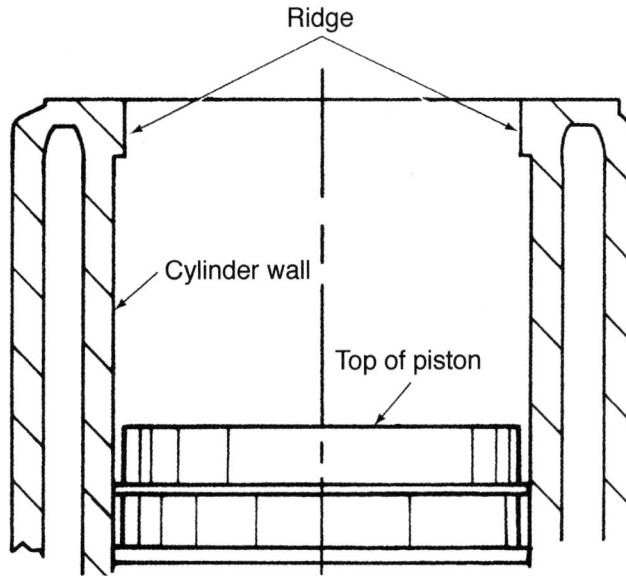

65. The figure above shows the cylinder and ring ridge. If the amount of cylinder wear does not require cylinder reboring, Technician A says the ring ridge at the top of the each cylinder can be removed with 400-grit sand paper. Technician B says the ring ridge at the top of the each cylinder can be removed with a 200-grit bead hone. Who is right?
    A. A only
    B. B only
    C. Both A and B
    D. Neither A nor B   (C.4)

66. While examining the old connecting rod bearings from an engine, the technician notices that the bearings from one rod are worn along the parting lines. This means that the technician should check carefully for:
    A. rod stretch.
    B. rod twisting.
    C. rod bending.
    D. a loose wrist pin.   (C.11)

67. When installing a piston/connecting rod into the cylinder block, which of the following steps is a technician LEAST to perform?
    A. Position the crankshaft journal at bottom dead center (BDC).
    B. Install boots over the rod bolts.
    C. Make sure the rings are installed right side up.
    D. Check piston-to-cylinder wall clearance using Plastigauge.   (C.12)

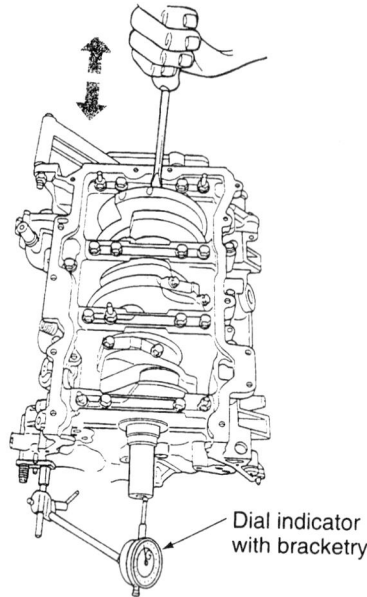

Dial indicator
with bracketry

68. In the figure above, Technician A says noise may be present if this check is not within specifications. Technician B says premature bearing wear could result if this check is not within specifications. Who is right?
    A. A only
    B. B only
    C. Both A and B
    D. Neither A nor B   (C.7)

69. The procedure for aligning cam and crankshaft sprockets before installing a timing chain varies from manufacturer to manufacturer. One step common to most procedures, however, is for the technician to:
    A. rotate the camshaft to fully open the intake valve in cylinder number 1.
    B. rotate the camshaft to fully open the exhaust valve in cylinder number 1.
    C. rotate the crankshaft to position piston number 1 at TDC.
    D. rotate the crankshaft to position piston number 1 at BDC.   (B.17)

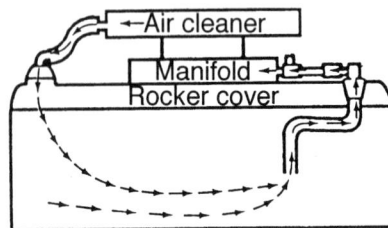

70. The figure above shows the positive crankcase ventilation (PCV) system. All of the following are symptoms of a stuck open PCV valve EXCEPT:
    A. blowby gases in the air filter.
    B. the engine stalling.
    C. rough idle operation.
    D. a lean air/fuel ratio.   (E.6)

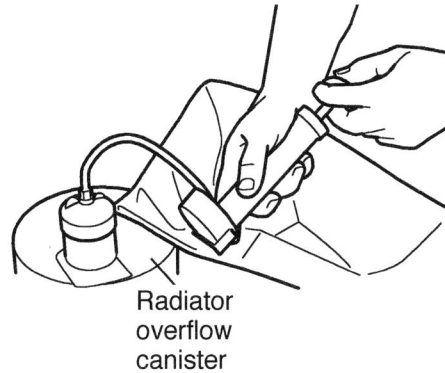

Radiator
overflow
canister

71. As shown above, a tester pump is being used to pressure test the engine cooling system. How much pressure should be applied to the cooling system when operating a tester pump?
   A. 15 psi (103 kPa)
   B. 17.5 psi (121 kPa)
   C. 20 psi (138 kPa)
   D. 22.5 psi (155 kPa)   (D.3)

72. A timing chain and sprockets are being installed in an OHV pushrod engine. Technician A says that the camshaft and crankshaft should be rotated so the chain and sprockets can be installed with the marks on both sprockets pointing straight up. Technician B says that the camshaft and crankshaft should be rotated so the chain and sprockets can be installed with the marks on both sprockets pointing straight down. Who is right?
   A. A only
   B. B only
   C. Both A and B
   D. Neither A nor B   (B.17)

73. If the specified valve seat angle is 45 degrees, many vehicle manufacturers recommend grinding the valve face to:
   A. 40 degrees.
   B. 50 degrees.
   C. 45.5 degrees.
   D. 44.5 degrees.   (B.7)

74. On turbocharged engines, the lubricating oil is frequently routed through an external oil cooler. The cooler prevents the oil from getting hot enough to oxide and thicken, a process that begins when oil reaches a temperature of:
   A. 100°F (38°C)
   B. 150°F (66°C)
   C. 200°F (93°C)
   D. 250°F (121°C)   (D.11)

75. The oil light on a vehicle stays on while the engine is running. Technician A say this could be caused by too much cam bearing clearance. Technician B says a grounded wire in the oil lamp warning lamp circuit could cause this. Who is right?
   A. A only
   B. B only
   C. Both A and B
   D. Neither A nor B   (D.1)

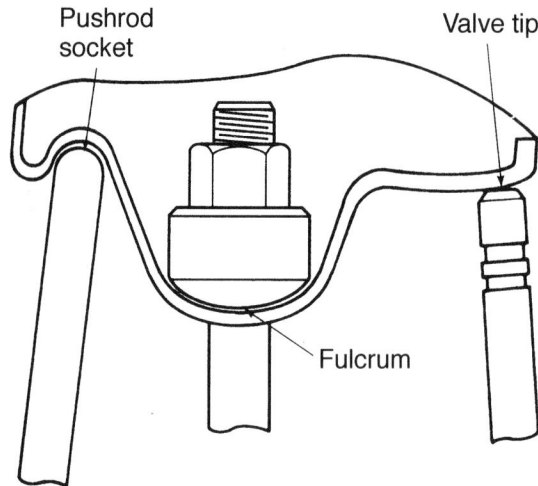

Pushrod socket   Valve tip

Fulcrum

76. The figure above shows an example of the rocker arm assembly on an engine equipped with hydraulic lifters. When adjusting valve lash on this engine while it is running, the step LEAST likely to be performed by a technician is:
    A. turning the adjusting nut clockwise 1/4 turn at a time.
    B. turning the adjusting nut clockwise 2 turns at a time.
    C. turning the adjusting nut counterclockwise until a clicking noise occurs.
    D. installing oil shrouds on the rocker arm.   (B.13)

77. Technician A says an exhaust manifold heat control valve stuck in the closed position causes a loss of engine power. Technician B says an exhaust manifold heat control valve stuck in the open position may cause an acceleration stumble. Who is right?
    A. A only
    B. B only
    C. Both A and B
    D. Neither A nor B   (E.8)

78. All of the following are reasons to replace a hydraulic valve lifter EXCEPT:
    A. excessive leak-down.
    B. convex bottom.
    C. pitted bottom.
    D. flat bottom.   (B.12)

79. A vehicle is equipped with a coolant recovery system. Coolant does not return to the radiator when the engine cools. Technician A says that the transfer hose may be plugged. Technician B says that the filler neck soldered joint could be cracked. Who is right?
    A. A only
    B. B only
    C. Both A and B
    D. Neither A nor B   (D.9)

80. The LEAST likely cause of an oil saturated PCV filter is:
    A. worn piston rings.
    B. an obstructed PCV vacuum hose.
    C. a stuck open PCV valve.
    D. a clogged PCV valve.   (E.6)

81. The customer says that the engine requires excessive cranking to start. The LEAST likely cause of this problem would be:
    A. a cracked cylinder block.
    B. a jumped timing belt.
    C. a faulty fuel pump.
    D. a stuck-open EGR valve.   (A.2)

82. A engine equipped with electronic fuel injection has a loose exhaust manifold. Technician A says that the loose manifold may cause noisy engine operation. Technician B says that the loose manifold may cause poor vehicle driveability. Who is right?
    A. A only
    B. B only
    C. Both A and B
    D. Neither A nor B   (E.8)

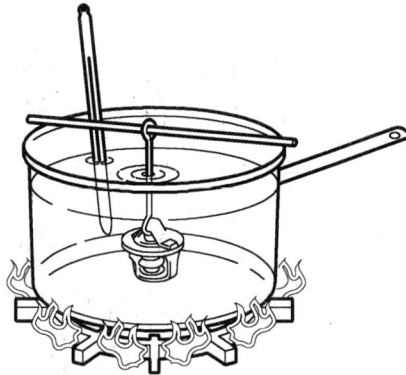

83. A thermostat-testing setup is shown in the figure above. Technician A says the thermostat will start to open when the water boils. Technician B says the thermostat valve should be fully open when the temperature equals the rated temperature stamped on the thermostat. Who is right?
    A. A only
    B. B only
    C. Both A and B
    D. Neither A nor B   (D.6)

84. Which of the following is LEAST likely to cause engine noise?
    A. Loose pistons
    B. Worn cylinders
    C. Worn main bearings
    D. Loose camshaft bearings   (A.4)

85. Serpentine belt stretch is indicated by:
    A. using a belt tension gauge.
    B. belt deflection.
    C. a squealing noise at idle.
    D. using the scale on the tensioner housing.   (D.4)

86. All of the following are true about torque-to-yield head bolts EXCEPT:
    A. they permanently stretch when they are torqued.
    B. they must be torqued in a proper sequence.
    C. many older engines have torque-to-yield head bolts.
    D. torque-to-yield head bolts are usually tightened to a specific torque and then rotated tighter.   (B.18)

87. Technician A says that an engine oil cooler can be located inside one of the radia-
tor tanks. Technician B says that an engine oil cooler can be mounted ahead of
the radiator support. Who is right?
A. A only
B. B only
C. Both A and B
D. Neither A nor B   (D.11)

88. A technician is testing an upper radiator hose by squeezing it. The most likely
cause of crackling or crunching noises would be:
A. a corroded anti-collapse spring.
B. low coolant level.
C. a deteriorated hose inner liner.
D. damage due to contact with power steering fluid.   (D.5)

89. Technician A says that air trapped in the cooling system can cause overheating
and a cracked cylinder head. Technician B says on some engines you can unscrew
the coolant temperature sender to bleed the air out of the system. Who is right?
A. A only
B. B only
C. Both A and B
D. Neither A nor B   (D.7)

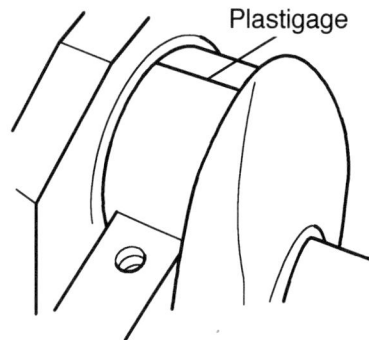
Plastigage

90. To measure bearing clearance, install a strip of Plastigage across the journal, as in
the figure above, and then tighten the bearing cap to the specified torque.
Remove the bearing cap and measure the width of the Plastigage on the journal
with which of the following?
A. The Plastigage package
B. A ruler
C. Dial calipers
D. A micrometer   (C.7)

91. A technician is using a dial indicator to measure valve-to-guide clearance. When
the valve head is moved from side to side, the dial indicator shows a maximum
value of 0.004 inch (0.102 mm). This means that valve guide clearance is:
A. 0.001 inch (0.025 mm).
B. 0.002 inch (0.051 mm).
C. 0.004 inch (0.102 mm).
D. 0.008 inch (0.203 mm).   (B.6)

92. The LEAST likely step in a diagnostic procedure would be to:
    A. question the customer for more information regarding the problem.
    B. be sure that the customer complaint is eliminated.
    C. start with the most difficult test.
    D. road test the vehicle.   (A.1)

93. A cylinder balance test on a carbureted engine has revealed one cylinder that is contributing much less power than the other cylinders. The LEAST likely cause of this problem is:
    A. a faulty ignition system.
    B. a burned exhaust valve.
    C. a faulty carburetor.
    D. a leaking intake manifold.   (A.7)

94. Technician A says that valve rotators should be disassembled and cleaned during an engine overhaul. Technician B says that a rotator causing the valve to rotate in either direction is functioning properly.
    A. A only
    B. B only
    C. Both A and B
    D. Neither A nor B   (B.4)

95. Technician A says that a press-fit harmonic balancer should be removed using a three-jaw gear puller. Technician B says that a harmonic balancer with a damaged keyway should be replaced. Who is right?
    A. A only
    B. B only
    C. Both A and B
    D. Neither A nor B   (C.13)

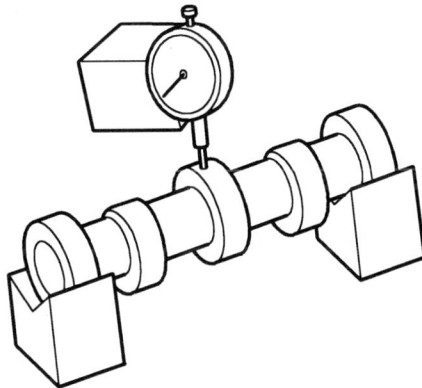

96. The measuring tool in the figure above is checking the camshaft:
    A. journal condition.
    B. runout.
    C. lift.
    D. bearing clearance.   (B.15)

97. An excessive sulfur smell in the exhaust of a vehicle with a catalytic converter can be an indication of:
    A. a lean fuel mixture.
    B. coolant leaking into a combustion chamber.
    C. a rich fuel mixture.
    D. a vacuum leak.   (A.5)

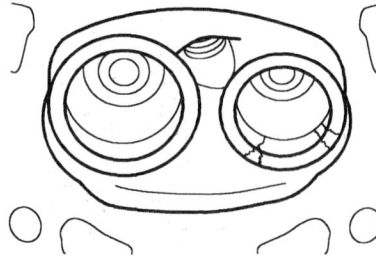

98. As seen in the figure above, a valve seat has been found to be cracked. Technician A says valve seat inserts may be removable. Technician B says a valve seat insert should be staked after installation. Who is right?
    A. A only
    B. B only
    C. Both A and B
    D. Neither A nor B   (B.8)

99. A low, steady vacuum gauge reading as shown above indicates:
    A. burned or leaking valves.
    B. a late ignition timing.
    C. weak valve springs.
    D. a leaking head gasket.   (A.6)

100. A gear type oil pump is being cleaned and inspected. Technician A says that gear thickness and backlash are not usually measured; the gears and housing are only inspected for scoring and damage. Technician B says that gear thickness and backlash should always be measured. Who is right?
    A. A only
    B. B only
    C. Both A and B
    D. Neither A nor B   (D.2)

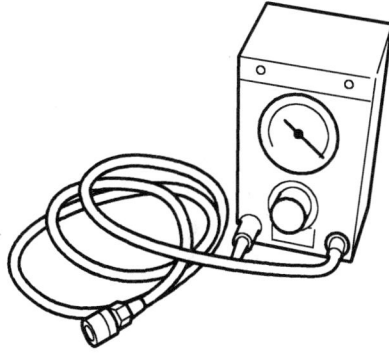

101. During a cylinder leakage test using a cylinder leakage tester, as shown in the figure above, the reading on the leakage tester exceeds 20 percent. The LEAST likely place that the technician would check for leaking air would be from the:
    A. tool air hose.
    B. vehicle tailpipe.
    C. radiator filler neck.
    D. PCV valve opening in the rocker arm cover.   (A.9)

102. Technician A says that filling a vehicle's cooling system with pure coolant provides the maximum protection against freezing. Technician B says that pure water absorbs heat better than pure coolant. Who is right?
    A. A only
    B. B only
    C. Both A and B
    D. Neither A nor B   (D.7)

103. Technician A says after the valve seats are resurfaced, install blue dye on the valve face and install the valve against the seat. Technician B says when the blue does not appear 360 degrees around the valve face, replace the valve. Who is right?
    A. A only
    B. B only
    C. Both A and B
    D. Neither A nor B   (B.9)

104. When blowing out an air filter element, how far should the gun be from the inside of the element?
    A. 5 inches (127 mm)
    B. 6 inches (152 mm)
    C. 7 inches (179 mm)
    D. 8 inches (203 mm)   (E.2)

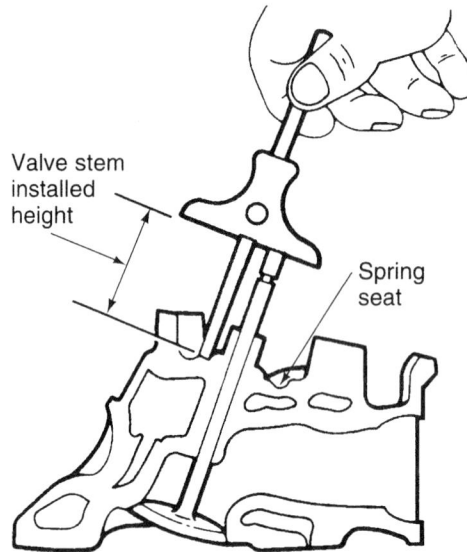

Valve stem installed height

Spring seat

105. In the above figure, Technician A says the tool is being used to check combustion chamber volume. Technician B says the tool is being used to check valve stem installed height. Who is right?
A. A only
B. B only
C. Both A and B
D. Neither A nor B   (B.10)

106. During a cylinder leakage test, air comes out the PCV valve opening in the rocker arm cover. This is an indication of:
A. worn intake valves.
B. worn exhaust valves.
C. a broken PCV valve.
D. worn piston rings.   (A.9)

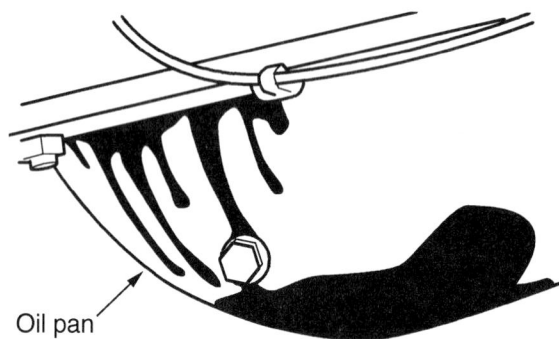

Oil pan

107. As shown in the figure above, even a small oil leak can result in major oil loss. It has been estimated that three drops of oil leaking every 100 feet results in a total of 3 quarts (2.8 liters) of oil loss every:
A. 100 miles (161 km).
B. 500 miles (804 km).
C. 1,500 miles (2,413 km).
D. 1,000 miles (1,609 km).   (A.3)

108. On some engines that use a timing chain hydraulic tensioner, chain stretch can be checked for by:
    A. using a chain tension gauge.
    B. measuring the feed oil pressure.
    C. measuring tensioner length.
    D. rotating the crankshaft backwards by hand while watching the distributor rotor.   (B.14)

109. Technician A says that when removing a cylinder head from an overhead camshaft (OHC) engine, the timing belt or chain will have to be removed from the block. Technician B says if the timing belt is to be reused, mark its direction of rotation and match it during reassembly. Who is right?
    A. A only
    B. B only
    C. Both A and B
    D. Neither A nor B   (B.1)

110. If the starter motor does not crank the engine, the first diagnostic step that the technician should perform is to:
    A. disable the ignition system.
    B. remove the spark plugs.
    C. rotate the engine by hand.
    D. watch for oil or coolant flow from the spark plug holes.   (A.2)

111. Technician A says blue-gray smoke coming from the exhaust may be caused by stuck piston rings. Technician B says this could be caused by a plugged oil drain passage in the cylinder head. Who is right?
    A. A only
    B. B only
    C. Both A and B
    D. Neither A nor B   (A.5)

112. Technician A says that positive type valve stem seals must be installed before the valves are installed in the cylinder head. Technician B says that positive type valve stem seals must be pushed down firmly over the top of the valve guides. Who is right?
    A. A only
    B. B only
    C. Both A and B
    D. Neither A nor B   (B.5)

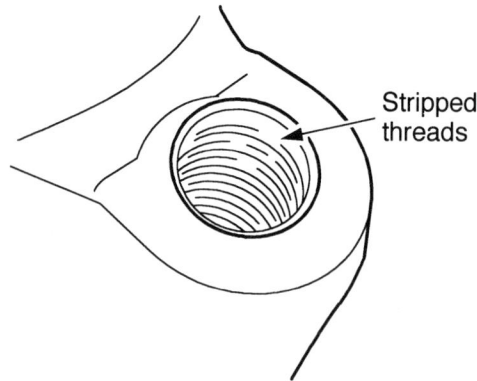

Stripped threads

113. If threads are damaged, as shown in the figure above, the opening may be drilled and threaded, and then a heli-coil may be installed to provide a thread:
    A. the same size as the new hole.
    B. one thread size smaller than the original.
    C. one thread size larger than the original.
    D. the same size as the original.   (C.3)

114. A cooling system is being checked with a pressure tester. The gauge pressure rises when the engine is started. Technician A says this could be caused by a crack in the combustion chamber. Technician B says this could be caused by clogged tubes in the radiator. Who is right?
    A. A only
    B. B only
    C. Both A and B
    D. Neither A nor B   (D.3)

115. Technician A says too much turbo boost can result in a damaged wastegate diaphragm. Technician B says too much turbo boost can result in bent compressor wheel blades. Who is right?
    A. A only
    B. B only
    C. Both A and B
    D. Neither A nor B   (E.3)

116. Technician A says that when installing an intake manifold that uses synthetic rubber seals at the front and rear ends, the top and bottom of the rubber seals should be coated with silicone sealer. Technician B says that only a dab of silicone sealer should be placed at the very ends of the seals. Who is right?
    A. A only
    B. B only
    C. Both A and B
    D. Neither A nor B   (E.1)

117. The LEAST likely cause of camshaft bind would be:
    A. excessive runout.
    B. an improperly installed bearing.
    C. bore misalignment.
    D. excessive bearing clearance.   (B.16)

118. A technician has added fluorescent dye to an engine crankcase in order to locate an oil leak. The dye will glow when it is exposed to:
    A. a fluorescent light.
    B. an ultraviolet light.
    C. a strobe light.
    D. an infrared light.   (A.3)

119. When installing a timing chain cover, the step that a technician is LEAST likely to perform would be:
    A. making sure the woodruff key is in place.
    B. making sure the oil slinger is in place.
    C. making sure the piston in cylinder #1 is at TDC.
    D. making sure the oil seal has been lubricated.   (C.15)

120. Technician A says that a damaged starter ring gear on a manual transmission flywheel can usually be replaced. Technician B says that a damaged starter ring gear on an automatic transmission flywheel can usually be replaced. Who is right?
    A. A only
    B. B only
    C. Both A and B
    D. Neither A nor B   (C.14)

121. Technician A says that carelessly installed valve stem seals may cause excessive oil consumption. Technician B says that the valve stems should be lubricated with grease before the valves are installed. Who is right?
    A. A only
    B. B only
    C. Both A and B
    D. Neither A nor B   (B.5)

122. While inspecting the intake manifold from a V-type engine, a technician notices a crack in the exhaust gas crossover passage on the underside of the manifold. The most likely cause of this condition is:
    A. the intake manifold bolts were overtorqued.
    B. the intake manifold bolts were not torqued in the correct sequence.
    C. the EGR passage in the manifold is plugged with carbon.
    D. the heat riser valve on one of the exhaust manifolds is stuck shut.   (E.1)

123. Technician A says that camshaft lobe lift can be checked with the camshaft still mounted in the engine. Technician B says that camshaft runout can be checked with the camshaft still mounted in the engine. Who is right?
    A. A only
    B. B only
    C. Both A and B
    D. Neither A nor B   (B.15)

124. While replacing a faulty heater core hose, a technician discovers that the hose is stubbornly stuck to the heater core fitting. To remove the hose, he should:
    A. pierce the hose with a screwdriver and try to pull it off the fitting.
    B. use a pair of pliers to loosen the hose.
    C. pry behind the edge of the hose to loosen it.
    D. cut the hose off at the fitting and slit the remaining piece.   (D.5)

7 Appendices

Answers to the Test Questions for the Sample Test Section 5

1.	A	17.	C	33.	A	49.	A
2.	D	18.	D	34.	A	50.	B
3.	A	19.	D	35.	A	51.	C
4.	D	20.	C	36.	A	52.	D
5.	D	21.	A	37.	D	53.	A
6.	C	22.	D	38.	A	54.	B
7.	C	23.	A	39.	A	55.	C
8.	B	24.	D	40.	C	56.	B
9.	D	25.	D	41.	C	57.	D
10.	A	26.	D	42.	B	58.	A
11.	A	27.	B	43.	C	59.	C
12.	B	28.	C	44.	B	60.	D
13.	C	29.	B	45.	B	61.	B
14.	B	30.	A	46.	D	62.	C
15.	C	31.	A	47.	D		
16.	D	32.	B	48.	D		

Explanations to the Answers for the Sample Test Section 5

Question #1
Answer A is correct.
Answer B is wrong. Air leaking from the tailpipe indicates a leaking exhaust valve.
Answer C is wrong. Only Technician A is right.
Answer D is wrong. Technician A is right.

Question #2
Answer A is wrong. The bores in the piston and the rod should be aligned before pressing in the wrist pin.
Answer B is wrong. The small end of the rod should be heated before pressing in the wrist pin.
Answer C is wrong. Position marks on the piston and rod should oriented properly before pressing in the wrist pin.
Answer D is correct.

Question #3
Answer A is correct.
Answer B is wrong. The vibration damper should be replaced if the inertia ring is loose.
Answer C is wrong. The vibration damper should be replaced if the rubber is cracked.
Answer D is wrong. The vibration damper should be replaced if the rubber is oil-soaked.

Question #4
Answer A is wrong. Air should be blown from the inside out to remove particles.
Answer B is wrong. Air forced into the filter at too short of a distance may damage the filter.
Answer C is wrong. Air forced into the filter at too short of a distance may damage the filter.
Answer D is correct.

Question #5
Answer A is wrong. The piston skirt would not be damaged if the ring ridge is not removed.
Answer B is wrong. The piston pin would not be broken if the ring ridge is not removed.
Answer C is wrong. Connecting rod bearings would not be affected by the piston ring ridge.
Answer D is correct.

Question #6
Answer A is wrong. Measurements A and B indicate vertical taper.
Answer B is wrong. Measurements C and D indicate horizontal taper.
Answer C is correct.
Answer D is wrong. Measurements A and D do not indicate out-of-round.

Question #7
Answer A is wrong. The tester may be used to test for cooling system leaks.
Answer B is wrong. The tester may be used to test the radiator cap pressure release valve.
Answer C is correct.
Answer D is wrong. The tester may be used to test for heater core leaks.

Question #8
Answer A is wrong. The belt would have to be severely loose or worn to cause a discharged battery.
Answer B is correct.
Answer C is wrong. The belt would have to be severely loose or worn to cause poor power steering assist.
Answer D is wrong. The belt would have to be severely loose or worn to cause engine overheating.

Question #9
Answer A is wrong. A manifold heat control valve improves fuel vaporization in the intake manifold.
Answer B is wrong. A manifold heat control valve stuck in the closed position causes a loss of engine power.
Answer C is wrong. A manifold heat control valve stuck in the open position may cause acceleration stumble.
Answer D is correct.

Question #10
Answer A is correct.
Answer B is wrong. Manifold vacuum is reduced during acceleration, therefore the leak would not contribute to cylinder misfire.
Answer C is wrong. Only Technician A is right.
Answer D is wrong. Technician A is right.

Question #11
Answer A is correct.
Answer B is wrong. Oil coolers do not relieve excessive oil pressure.
Answer C is wrong. Oil coolers do not prevent oil pump wear.
Answer D is wrong. Oil coolers do not prevent main bearing wear.

Question #12
Answer A is wrong. Excessive valve spring height may result in float or cylinder misfiring.
Answer B is correct.
Answer C is wrong. Only Technician B is right.
Answer D is wrong. Technician B is right.

Question #13
Answer A is wrong.
Answer B is wrong.
Answer C is correct. Both sticking valves and improper valve timing can cause bent pushrods.
Answer D is wrong.

Question #14
Answer A is wrong. Pump cover flatness may be measured with a feeler gauge.
Answer B is correct.
Answer C is wrong. The clearance between the inner and outer rotors may be measured with a feeler gauge while they are still installed.
Answer D is wrong. The clearance between the top of the rotors and a straightedge may be measured with a feeler gauge.

Question #15
Answer A is wrong.
Answer B is wrong.
Answer C is correct. Bolting a cylinder head to a warped cylinder block deck may cause the head to bend. This could cause valve seat distortion. The head gasket will not be uniformly compressed, which can lead to coolant and combustion leaks.
Answer D is wrong.

Question #16
Answer A is wrong. The first step is to drill the opening to allow heli-coil installation.
Answer B is wrong. A heli-coil is not installed with a drill.
Answer C is wrong. Both Technicians A and B are right.
Answer D is correct.

Question #17
Answer A is wrong.
Answer B is wrong.
Answer C is correct. When measuring valve stem-to-guide clearance, stem diameter and guide diameter should be measured at three vertical locations: near the top of the guide, the middle of the guide, and the bottom of the guide. Either a hole gauge or a snap gauge can be used to measure the guide diameter.
Answer D is wrong.

Question #18
Answer A is wrong. This temperature is not hot enough to clean the iron and steel.
Answer B is wrong. The lower range is too low to clean iron and steel.
Answer C is wrong. The upper range is too hot and may begin to cause damage to the parts.
Answer D is correct.

Question #19
Answer A is wrong. Worn camshaft bearings may cause low oil pressure.
Answer B is wrong. Worn crankshaft bearings may cause low oil pressure.
Answer C is wrong. Weak oil pressure regulator spring tension may cause low oil pressure.
Answer D is correct.

Question #20
Answer A is wrong.
Answer B is wrong.
Answer C is correct. Burrs on the crankshaft flange might prevent the flywheel or flexplate from seating properly. During engine operation, this will cause the flywheel or flexplate to wobble, damaging the flywheel and starter gear teeth (during cranking) and the transmission torque converter bushing.
Answer D is wrong.

Question #21
Answer A is correct.
Answer B is wrong. A bent connecting rod will not cause uneven main bearing wear.
Answer C is wrong. A bent connecting rod will not cause uneven piston pin wear.
Answer D is wrong. A bent connecting rod will not cause excessive cam bearing wear.

Question #22
Answer A is wrong. On some engines, the crankshaft front oil seal can be replaced without removing the timing cover.
Answer B is wrong. When installing a lip seal, the lip must face toward the fluid being sealed.
Answer C is wrong. Both Technicians A and B are wrong.
Answer D is correct.

Question #23
Answer A is correct.
Answer B is wrong. If the cylinder had a burned exhaust valve, compression would not increase during a wet test.
Answer C is wrong. If the cylinder had a bent intake valve, compression would not increase during a wet test.
Answer D is wrong. If the engine had a worn camshaft lobe, compression would not increase during a wet test.

Question #24
Answer A is wrong. The cooling fan would only operate when the condenser switch is closed.
Answer B is wrong. The cooling fan may operate when the condenser switch is closed.
Answer C is wrong. The cooling fan motor would most likely burn out as a result of a short to ground.
Answer D is correct.

Question #25
Answer A is wrong. RTV sealant is likely to be used when installing an oil pan.
Answer B is wrong. RTV sealant is likely to be used when installing a valve cover.
Answer C is wrong. RTV is likely to be used when installing an intake manifold.
Answer D is correct.

Question #26
Answer A is wrong. A damaged sealing gasket would not cause a collapsed hose.
Answer B is wrong. A damaged seat would not cause a collapsed hose.
Answer C is wrong. A damaged expansion tank would not cause a collapsed hose.
Answer D is correct.

Question #27
Answer A is wrong. The coolant boiling point rises as the pressure is increased.
Answer B is correct.
Answer C is wrong. Only Technician B is right.
Answer D is wrong. Technician B is right.

Question #28
Answer A is wrong.
Answer B is wrong.
Answer C is correct. A defective water pump bearing may cause a growling noise when the engine is idling, and the water pump bearing may be ruined by coolant leaking past the pump seal.
Answer D is wrong.

Question #29
Answer A is wrong. Worn valve stem seals generally do not cause stem or guide wear.
Answer B is correct.
Answer C is wrong. Only Technician B is right.
Answer D is wrong. Technician B is right.

Question #30
Answer A is correct.
Answer B is wrong. Valve spring squareness is not checked by rolling the spring on a surface plate.
Answer C is wrong. Only Technician A is right.
Answer D is wrong. Technician A is right.

Question #31
Answer A is correct.
Answer B is wrong. Worn valve lock grooves will not cause a clicking noise at idle.
Answer C is wrong. Only Technician A is right.
Answer D is wrong. Technician A is right.

Question #32
Answer A is wrong. The piston must be at TDC on the compression stroke.
Answer B is correct.
Answer C is wrong. Only Technician B is right.
Answer D is wrong. Technician B is right.

Question #33
Answer A is correct.
Answer B is wrong. This measurement exceeds specification.
Answer C is wrong. This measurement exceeds specification.
Answer D is wrong. This measurement exceeds specification.

Question #34
Answer A is correct.
Answer B is wrong. A pry bar may be used to bottom the crankshaft.
Answer C is wrong. A dial indicator may be used to take measurements.
Answer D is wrong. Feeler gauges may be used to take measurements.

Question #35
Answer A is correct.
Answer B is wrong. If the engine is cranking properly, the battery would have sufficient voltage.
Answer C is wrong. Compression would not be the first test to perform.
Answer D is wrong. Because engine vacuum is low during cranking, a vacuum test would not be conclusive.

Question #36
Answer A is correct.
Answer B is wrong. The thermostat should be installed so that the thermal element points toward the engine.
Answer C is wrong. Only Technician A is right.
Answer D is wrong. Technician A is right.

Question #37
Answer A is wrong. The valve lifter height would not be measured in this manner.
Answer B is wrong. Camshaft journal out-of-round should be measured with a micrometer.
Answer C is wrong. Pushrod length would not be measured in this manner.
Answer D is correct.

Question #38
Answer A is correct.
Answer B is wrong. Excessive camshaft bearing clearance will not cause a clicking noise at idle.
Answer C is wrong. Only Technician A is right.
Answer D is wrong. Technician A is right.

Question #39
Answer A is correct.
Answer B is wrong. Camshaft gear backlash cannot be measured with a micrometer.
Answer C is wrong. Only Technician A is right.
Answer D is wrong. Technician A is right.

Question #40
Answer A is wrong.
Answer B is wrong.
Answer C is correct. Both a misfiring ignition system and an intake manifold leak could cause a cylinder to contribute too little power.
Answer D is wrong.

Question #41
Answer A is wrong.
Answer B is wrong.
Answer C is correct. Improper valve timing may cause reduced engine power and, on some engines, bent valves.
Answer D is wrong.

Question #42
Answer A is wrong. Lifter bottoms must be convex.
Answer B is correct.
Answer C is wrong. Only Technician B is right.
Answer D is wrong. Technician B is right.

Question #43
Answer A is wrong. The rocker arm assembly would not need to be removed first when performing timing belt removal on an OHC engine.
Answer B is wrong. The position of the timing belt is not vital to installation.
Answer C is correct.
Answer D is wrong. The water pump would not need to be removed when performing timing belt removal on an OHC engine.

Question #44
Answer A is wrong.
Answer B is correct. Warpage on the camshaft side of the cylinder head must be measured before determining whether the head can be surfaced and reused.
Answer C is wrong.
Answer D is wrong.

Question #45
Answer A is wrong. A PCV valve stuck open may cause acceleration stumble.
Answer B is correct.
Answer C is wrong. A restricted PCV hose would not cause the engine to surge at high speed.
Answer D is wrong. A restricted PCV hose would not lead to engine detonation.

Question #46
Answer A is wrong. The tool shown does not measure valve guide depth.
Answer B is wrong. The tool shown does not measure valve seat angle.
Answer C is wrong. A straightedge and feeler gauge are used to measure cylinder head flatness.
Answer D is correct.

Question #47
Answer A is wrong. Excessive heat will cause a heater hose to become hard and brittle.
Answer B is wrong. Engine oil will cause a radiator hose to become soft and gummy.
Answer C is wrong. Both Technicians A and B are wrong.
Answer D is correct.

Question #48
Answer A is wrong. Adding shims will increase the clearance.
Answer B is wrong. Adding shims will increase the clearance.
Answer C is wrong. Adding a 0.015 inch (0.381 mm) shim will increase clearance by 0.005 inch (0.127 mm).
Answer D is correct.

Question #49
Answer A is correct. Piston ring clearance is measured by placing a new ring in the groove and then inserting a feeler gauge between the ring and the groove.
Answer B is wrong.
Answer C is wrong.
Answer D is wrong.

Question #50
Answer A is wrong. A wastegate valve stuck closed would cause increased boost pressure.
Answer B is correct.
Answer C is wrong. A leaking wastegate diaphragm may cause increased boost pressure.
Answer D is wrong. A disconnected wastegate linkage would cause increased boost pressure because the valve would never open.

Question #51
Answer A is wrong.
Answer B is wrong.
Answer C is correct. Torque-to-yield bolts provide a more uniform clamping force than conventional bolts. They are often tightened by being torqued to a specific lb-ft value, and then rotated a specific number of degrees.
Answer D is wrong.

Question #52
Answer A is wrong. The tool can also be used to install camshaft bearings.
Answer B is wrong. The tool can also be used to remove camshaft bearings.
Answer C is wrong. The tool is not used for taking measurements.
Answer D is correct.

Question #53
Answer A is correct.
Answer B is wrong. Balance shafts should be timed in relation to the crankshaft.
Answer C is wrong. Only Technician A is right.
Answer D is wrong. Technician A is right.

Question #54
Answer A is wrong. The easier diagnostic procedures should be performed before attempting more difficult procedures.
Answer B is correct.
Answer C is wrong. Only Technician B is right.
Answer D is wrong. Technician B is right.

Question #55
Answer A is wrong. The vacuum advance does control spark advance in relation to engine load.
Answer B is wrong. The mechanical advance does control spark advance in relation to engine speed.
Answer C is correct.
Answer D is wrong. The vacuum advance does rotate the pickup plate in the opposite direction to shaft rotation.

Question #56
Answer A is wrong. A margin that is less than 1/64 inch (0.4 mm) will not cause a clicking noise at idle.
Answer B is correct.
Answer C is wrong. The angle of the cut, not the margin, affects valve seating.
Answer D is wrong. The margin shown will not cause valve seat recession.

Question #57
Answer A is wrong. Late ignition timing would result in a low, steady reading.
Answer B is wrong. Intake manifold leaks would cause a very low, steady reading.
Answer C is wrong. A restricted exhaust system would cause vacuum to slowly decrease after the engine was accelerated and held steady.
Answer D is correct.

Question #58
Answer A is correct.
Answer B is wrong. The battery should recover the lost voltage over a short period of no demand.
Answer C is wrong. The battery passed the load test.
Answer D is wrong. The battery passed the load test.

Question #59
Answer A is wrong.
Answer B is wrong.
Answer C is correct. Oil leaking from the crankshaft rear main bearing seal could be caused by a faulty oil seal or a malfunctioning PCV system.
Answer D is wrong.

Question #60
Answer A is wrong. An intake manifold leak causes a high-pitched whistle at idle and at low speeds.
Answer B is wrong. A choke stuck closed will not cause a high-pitched whistle.
Answer C is wrong. A fuel system leak normally would not cause a noise.
Answer D is correct.

Question #61
Answer A is wrong. Worn pistons and cylinders would cause a thumping noise during acceleration.
Answer B is correct.
Answer C is wrong. Worn main bearings cause a thump when the engine is started.
Answer D is wrong. Loose camshaft bearings would cause a growling noise at all times.

Question #62
Answer A is wrong.
Answer B is wrong.
Answer C is correct. Removable valve seat inserts may be removed using a special puller or a pry bar. Also, a special driver is used to install a new valve seat insert and the insert should be staked in place.
Answer D is wrong.

Answers to the Test Questions for the Additional Test Questions Section 6

1.	B	32.	D	63.	D	94.	B
2.	A	33.	D	64.	A	95.	B
3.	C	34.	D	65.	D	96.	B
4.	C	35.	B	66.	A	97.	C
5.	A	36.	D	67.	D	98.	C
6.	D	37.	C	68.	C	99.	B
7.	A	38.	A	69.	C	100.	A
8.	C	39.	C	70.	A	101.	A
9.	C	40.	C	71.	A	102.	B
10.	C	41.	A	72.	D	103.	C
11.	D	42.	B	73.	D	104.	B
12.	D	43.	D	74.	D	105.	B
13.	B	44.	C	75.	C	106.	D
14.	C	45.	B	76.	B	107.	D
15.	A	46.	B	77.	C	108.	C
16.	B	47.	C	78.	B	109.	B
17.	B	48.	D	79.	C	110.	A
18.	B	49.	C	80.	C	111.	C
19.	A	50.	D	81.	A	112.	B
20.	D	51.	A	82.	C	113.	D
21.	D	52.	B	83.	D	114.	A
22.	D	53.	A	84.	D	115.	D
23.	C	54.	C	85.	D	116.	B
24.	D	55.	A	86.	C	117.	D
25.	D	56.	B	87.	C	118.	B
26.	A	57.	B	88.	C	119.	C
27.	A	58.	B	89.	C	120.	A
28.	C	59.	A	90.	A	121.	A
29.	D	60.	C	91.	B	122.	D
30.	D	61.	A	92.	C	123.	A
31.	C	62.	A	93.	C	124.	D

Explanations to the Answers for the Additional Test Questions Section 6

Question #1
Answer A is wrong. Valve guide height is generally measured from the top of the spring seat to the top of the guide.
Answer B is correct.
Answer C is wrong. Only Technician B is right.
Answer D is wrong. Technician B is right.

Question #2
Answer A is correct. A PCV valve stuck in the open position would not cause blue exhaust smoke.
Answer B is wrong.
Answer C is wrong.
Answer D is wrong.

Question #3
Answer A is wrong.
Answer B is wrong.
Answer C is correct. It is especially important to thoroughly clean engine oil passages after bearing failure. It is also important that other parts of the lubrication system which may contain metal particles be replaced to prevent future engine damage.
Answer D is wrong.

Question #4
Answer A is wrong. A variance of 0.005 inch (0.127 mm) is within specifications.
Answer B is wrong. A variance of 0.0625 inch (1.588 mm) is within specifications.
Answer C is correct. A variance of 0.125 inch (3.175 mm) is beyond specifications.
Answer D is wrong. A variance of 0.025 inch (0.635 mm) is within specifications.

Question #5
Answer A is correct.
Answer B is wrong. A broken timing chain could cause a bent pushrod.
Answer C is wrong. A sticking valve could cause a bent pushrod.
Answer D is wrong. Improper valve adjustment could cause a bent pushrod.

Question #6
Answer A is wrong. RTV sealant is never used on threaded fasteners.
Answer B is wrong. Anaerobic sealant fumes will not harm an O2 sensor.
Answer C is wrong. Both Technicians A and B are wrong.
Answer D is correct.

Question #7
Answer A is correct.
Answer B is wrong. Plastigage cannot effectively measure bearing alignment.
Answer C is wrong. A dial indicator could not measure bearing alignment.
Answer D is wrong. A telescoping gauge is not used to measure bearing alignment.

Question #8
Answer A is wrong.
Answer B is wrong.
Answer C is correct. Balance shafts should be checked for runout following the same procedure used to check camshafts for runout. Their journals should be checked for taper following the same procedure used to check crankshaft journals for taper.
Answer D is wrong.

Question #9
Answer A is wrong.
Answer B is wrong.
Answer C is correct. A corroded bearing in the water pump would cause the pump to be noisy. The cause of the corroded bearing could be a defective water pump seal.
Answer D is wrong.

Question #10
Answer A is wrong.
Answer B is wrong.
Answer C is correct. Both a bad ground in the cooling fan circuit and a bad wire to the fan relay could prevent the cooling fan from operating.
Answer D is wrong.

Question #11
Answer A is wrong. The mechanical advance weights rotate the cam or reluctor in the direction of distributor shaft rotation.
Answer B is wrong. The vacuum advance rotates the pickup or breaker plate in a direction opposite that of distributor shaft rotation.
Answer C is wrong. Both Technicians A and B are wrong.
Answer D is correct.

Question #12
Answer A is wrong. The cylinder head must be replaced.
Answer B is wrong. When the clearance is excessive, the cylinder head must be replaced.
Answer C is wrong. Inserting bushings is not recommended. The cylinder head must be replaced.
Answer D is correct.

Question #13
Answer A is wrong.
Answer B is correct. Excessive clearance between the starter pinion gear and the flywheel ring gear may cause a loud whining noise.
Answer C is wrong.
Answer D is wrong.

Question #14
Answer A is wrong. 1,500 rpm is too low for the test.
Answer B is wrong. 2,000 rpm is too low for the test.
Answer C is correct.
Answer D is wrong. 3,000 rpm is too high for the test.

Question #15
Answer A is correct.
Answer B is wrong. On some engines equipped with a distributor, ignition timing is not adjustable by rotating the distributor.
Answer C is wrong. Only Technician A is right.
Answer D is wrong. Technician A is right.

Question #16
Answer A is wrong. Ring grooves should be cleaned using a ring groove cleaning tool.
Answer B is correct.
Answer C is wrong. Only Technician B is right.
Answer D is wrong. Technician B is right.

Question #17
Answer A is wrong. Shims are not used in all vehicles.
Answer B is correct.
Answer C is wrong. Only Technician B is right.
Answer D is wrong. Technician B is right.

Question #18
Answer A is wrong. Installing a higher rated thermostat will not cause the engine to warm up faster. It will, however, cause it to operate at a higher temperature.
Answer B is correct.
Answer C is wrong. Only Technician B is right.
Answer D is wrong. Technician B is right.

Question #19
Answer A is correct.
Answer B is wrong. Engine coolant may boil due to low pressure in the system.
Answer C is wrong. Coolant may overflow through the damaged seal or gasket.
Answer D is wrong. If the pressure gets too low or enough coolant is lost, the engine may overheat.

Question #20
Answer A is wrong. This is not a three-angle valve job.
Answer B is wrong. This is not poor contact. The contact shown is correct.
Answer C is wrong. Both Technicians A and B are wrong.
Answer D is correct.

Question #21
Answer A is wrong. Low oil pressure would result in a continuous noise.
Answer B is wrong. Low oil level would result in a continuous noise.
Answer C is wrong. A worn lifter bottom would result in continuous noise.
Answer D is correct.

Question #22
Answer A is wrong. A low reading on two adjacent cylinders may indicate a blown head gasket.
Answer B is wrong. Carbon buildup would cause a high reading.
Answer C is wrong. A low reading on two adjacent cylinders may indicate a cracked cylinder head.
Answer D is correct.

Question #23
Answer A is wrong.
Answer B is wrong.
Answer C is correct. An overtensioned V-belt can damage an alternator front bearing. It can also cause the upper half of the crankshaft front main bearing to wear prematurely.
Answer D is wrong.

Question #24
Answer A is wrong. Always refer to manufacturers' recommendations.
Answer B is wrong. Always refer to manufacturers' recommendations.
Answer C is wrong. Always refer to manufacturers' recommendations.
Answer D is correct.

Question #25
Answer A is wrong. The cylinder head should not be used as a measurement location.
Answer B is wrong. The measurement should be to the top of the spring seat, not the top shim.
Answer C is wrong. The bottom of the shim should not be used as a measurement location.
Answer D is correct.

Question #26
Answer A is correct. The technician would not run a bottoming tap through the oil gallery bores because they have tapered pipe threads.
Answer B is wrong.
Answer C is wrong.
Answer D is wrong.

Question #27
Answer A is correct.
Answer B is wrong. The reading should be multiplied by two.
Answer C is wrong. The reading should be multiplied by two.
Answer D is wrong. The reading should be multiplied by two.

Question #28
Answer A is wrong. A dye penetrant must be used on aluminum heads.
Answer B is wrong. An electromagnetic-type tester is not used on pistons.
Answer C is correct.
Answer D is wrong. Aluminum intake manifold must be checked using a dye penetrant.

Question #29
Answer A is wrong. Shims are not added to adjust lash.
Answer B is wrong. Shims are not added to adjust lash.
Answer C is wrong. Valve adjustment is required.
Answer D is correct.

Question #30
Answer A is wrong. Excessive taper may require crankshaft grinding.
Answer B is wrong. Out-of-round journals may require crankshaft grinding.
Answer C is wrong. Excessive journal scoring may be removed through crankshaft grinding.
Answer D is correct.

Question #31
Answer A is wrong.
Answer B is wrong.
Answer C is correct. Both the valve spring and the valve stem seal can be replaced without removing the cylinder head from the engine.
Answer D is wrong.

Question #32
Answer A is wrong. The readings indicated that the battery is discharged. A discharged battery should not be load tested.
Answer B is wrong. The specific gravity readings do not vary enough between cylinders to warrant battery replacement. The battery should be charged and retested.
Answer C is wrong. The battery should not be refilled with fresh electrolyte; it should be recharged.
Answer D is correct.

Question #33
Answer A is wrong.
Answer B is wrong.
Answer C is wrong.
Answer D is correct. Camshaft bearings need not be positioned at the front of their bores or the rear of their bores. It most important that they be positioned to align the oil hole in the bearing with the oil supply passage in the bore.

Question #34
Answer A is wrong. Some balance shafts rotate at twice crankshaft speed.
Answer B is wrong. Balance shafts are commonly found on four-cylinder and V-type six-cylinder engines.
Answer C is wrong. Both Technicians A and B are wrong.
Answer D is correct.

Question #35
Answer A is wrong. Thermal cleaners heat parts from 650 to 800°F (343 to 427°C).
Answer B is correct.
Answer C is wrong. Only Technician B is right.
Answer D is correct. Technician B is right.

Question #36
Answer A is wrong. Disabling the faulty cylinder will cause engine rpm to drop less than for the other cylinders.
Answer B is wrong. Disabling the faulty cylinder will not cause the engine to stall.
Answer C is wrong. Both Technicians A and B are wrong.
Answer D is correct.

Question #37
Answer A is wrong.
Answer B is wrong.
Answer C is correct. When all the lifters in an engine are cupped, the camshaft and the lifters should be replaced.
Answer D is wrong.

Question #38
Answer A is correct.
Answer B is wrong. 20 seconds is too long.
Answer C is wrong. 25 seconds is too long.
Answer D is wrong. 30 seconds is too long.

Question #39
Answer A is wrong. Cam timing is not adjusted with a ruler.
Answer B is wrong. TDC is not located in this manner.
Answer C is correct.
Answer D is wrong. Valve lash is not adjusted at the timing chain.

Question #40
Answer A is wrong. Spring tension is not measured in this manner.
Answer B is wrong. Spring free height is not measured in this manner.
Answer C is correct. The block deck is checked for warpage using a straightedge and a feeler gauge. Also, minor nicks and burrs can be removed from the deck using a whetstone or a file.
Answer D is wrong.

Question #41
Answer A is correct. In general, a four-cylinder head can be warped up to 0.004 inch and still be within specifications.
Answer B is wrong.
Answer C is wrong.
Answer D is wrong.

Question #42
Answer A is wrong. A loose timing chain would not affect just two cylinders.
Answer B is correct.
Answer C is wrong. Only Technician B is right.
Answer D is wrong. Technician B is right.

Question #43
Answer A is wrong. Retarded timing would not result in gauge fluctuation.
Answer B is wrong. Advanced timing would not result in gauge fluctuation.
Answer C is wrong. A stuck EGR valve would not result in gauge fluctuation.
Answer D is correct.

Question #44
Answer A is wrong. Residue at the water pump drain hole may indicate a damaged seal or bearings.
Answer B is wrong. Coolant leaking from the water pump is a sign of failure.
Answer C is correct.
Answer D is wrong. A pressure tester should indicate a water pump leak.

Question #45
Answer A is wrong. RTV sealer dries in the presence of air by absorbing moisture from the air.
Answer B is correct.
Answer C is wrong. Only Technician B is right.
Answer D is wrong. Technician B is right.

Question #46
Answer A is wrong. If the switch is stuck closed, the fan motor will continue to run because the relay receives voltage directly from the battery.
Answer B is correct.
Answer C is wrong. Only Technician B is right.
Answer D is correct. Technician B is right.

Question #47
Answer A is wrong.
Answer B is wrong.
Answer C is correct. Torque-to-yield head bolts do stretch permanently when they are tightened, and some torque-to-yield bolts cannot be reused.
Answer D is wrong.

Question #48
Answer A is wrong. The use of unleaded fuel has nothing to do with interference angles.
Answer B is wrong. An interference angle will not result in a poor seal.
Answer C is wrong. Both Technicians A and B are wrong.
Answer D is correct.

Question #49
Answer A is wrong.
Answer B is wrong.
Answer C is correct. Main bearing bore out-of-round measurements of less than 0.001 inch are acceptable, so long as the vertical measurement is not the largest. The vertical measurement should never be the largest.
Answer D is wrong.

Question #50
Answer A is wrong. Stretched main bearing bores cannot be corrected by filing the caps.
Answer B is wrong. Stretched main bearing bores cannot be corrected by replacing the caps.
Answer C is wrong. Both Technicians A and B are wrong.
Answer D is correct.

Question #51
Answer A is correct. It is unlikely that the technician will measure center-to-center length of the connecting rods.
Answer B is wrong.
Answer C is wrong.
Answer D is wrong.

Question #52
Answer A is wrong.
Answer B is correct. The second measurement should be taken just below the ring ridge.
Answer C is wrong.
Answer D is wrong.

Question #53
Answer A is correct. Deck warpage less that 0.005 inch (0.127 mm) is usually acceptable.
Answer A is wrong.
Answer C is wrong.
Answer D is wrong.

Question #54
Answer A is wrong.
Answer B is wrong.
Answer C is correct. Some engines are fitted with a type of oil pump that fits over the crankshaft snout. Others, especially OHC belt-timed engines, have a belt driven pump.
Answer D is wrong.

Question #55
Answer A is correct.
Answer B is wrong. The oil rings should be installed first, then the second compression ring, and then the top compression ring.
Answer C is wrong. Only Technician A is right.
Answer D is wrong. Technician A is right.

Question #56
Answer A is wrong. A 0.250 inch (6.35 mm) lift cam and 1.5:1 rocker arms will cause the valve to open 0.375 inch (9.53 mm).
Answer B is correct.
Answer C is wrong. The rocker arm ratio does not determine which type of lifters must be used.
Answer D is wrong. The rocker arm ratio does not determine which type of lifters must be used.

Question #57
Answer A is wrong. A concentricity tester is a useful tool for measuring valve face and seat concentricity.
Answer B is correct.
Answer C is wrong. Blue dye can indicate if a valve has concentricity.
Answer D is wrong. Dial indicators on testers can check concentricity of the valve face and seat.

Question #58
Answer A is wrong.
Answer B is correct. Hammering the dished areas flat again is the best way to solve this common problem.
Answer C is wrong.
Answer D is wrong.

Question #59
Answer A is correct.
Answer B is wrong. Valves having rounded or uneven stem shoulders are replaced, not machined.
Answer C is wrong. Only Technician A is right.
Answer D is wrong. Technician A is right.

Question #60
Answer A is wrong.
Answer B is wrong.
Answer C is correct. Most manufacturers recommend that piston diameter be measured about 3/4 inch below the centerline of the wrist pin bore.
Answer D is wrong.

Question #61
Answer A is correct.
Answer B is wrong. Out-of-round is not measured in this way.
Answer C is wrong. Wear is not measured in this way.
Answer D is wrong. Bearing size is not measured in this way.

Question #62
Answer A is correct. Flywheel runout should be checked with the flywheel mounted on the crankshaft and the dial indicator mounted on the block or clutch cover so that it contacts the flywheel wear surface.
Answer B is wrong.
Answer C is wrong.
Answer D is wrong.

Question #63
Answer A is wrong. Thinking of possible causes should come after customer input.
Answer B is wrong. The customer should finish completely before questioning begins.
Answer C is wrong. A road test should be performed after the customer is heard.
Answer D is correct.

Question #64
Answer A is correct.
Answer B is wrong. Vibration damper rubber should be inspected for looseness.
Answer C is wrong. Vibration damper rubber should be inspected for cracking.
Answer D is wrong. Vibration damper rubber should be inspected for oil soaking.

Question #65
Answer A is wrong. A ridge reamer should be used to remove the ridge.
Answer B is wrong. Sandpaper should not be used to remove the ring ridge.
Answer C is wrong. Both Technicians A and B are wrong.
Answer D is correct.

Question #66
Answer A is correct. Bearings worn along the parting lines indicate rod stretch has occurred.
Answer B is wrong.
Answer C is wrong.
Answer D is wrong.

Question #67
Answer A is wrong. The crankshaft needs to be at BDC.
Answer B is wrong. Boots must be installed on the rod bolts to prevent damage to the crankshaft.
Answer C is wrong. Piston rings must be installed correct side up.
Answer D is correct.

Question #68
Answer A is wrong.
Answer B is wrong.
Answer C is correct. If crankshaft end play is excessive, a "clunk" noise may occur when the vehicle accelerates from a stop. Also, excessive back-and-forth motion of the crankshaft may cause rod and main bearings to wear out prematurely.
Answer D is wrong.

Question #69
Answer A is wrong.
Answer B is wrong.
Answer C is correct. Most timing belt installation procedures call for the crankshaft to be rotated until piston number 1 is at TDC.
Answer D is wrong.

Question #70
Answer A is correct.
Answer B is wrong. Engine stalling may be caused by a PCV valve stuck in the open position.
Answer C is wrong. Rough idle may occur if a PCV valve is stuck in the open position.
Answer D is wrong. A lean air/fuel mixture will result from a PCV valve stuck in the open position.

Question #71
Answer A is correct.
Answer B is wrong. This amount of pressure is too high.
Answer C is wrong. This amount of pressure is too high.
Answer D is wrong. This amount of pressure is too high.

Question #72
Answer A is wrong.
Answer B is wrong.
Answer C is wrong.
Answer D is correct. In most cases, the camshaft and crankshaft must be rotated so that the chain and sprockets can be installed with the marks on the sprockets pointing inward, toward each other.

Question #73
Answer A is wrong. 40 degrees is too acute.
Answer B is wrong. 50 degrees is too obtuse.
Answer C is wrong. 45.5 degrees is too obtuse.
Answer D is correct.

Question #74
Answer A is wrong.
Answer B is wrong.
Answer C is wrong.
Answer D is correct. Engine oil begins to oxidize and thicken when it reaches 250°F (121°C)

Question #75
Answer A is wrong.
Answer B is wrong.
Answer C is correct. Excessive cam bearing clearance or a grounded warning indicator circuit could cause the oil pressure light to remain on while the engine is running.
Answer D is wrong.

Question #76
Answer A is wrong.
Answer B is correct. The technician is unlikely to rotate the adjusting nut CW two turns at a time since this could cause the valve and piston to collide, damaging valve train components.
Answer C is wrong.
Answer D is wrong.

Question #77
Answer A is wrong.
Answer B is wrong.
Answer C is correct. A manifold heat control valve that is stuck closed may reduce engine power output. A valve that is stuck open may cause an acceleration stumble.
Answer D is wrong.

Question #78
Answer A is wrong. Valve lifters should be checked for leak-down.
Answer B is correct.
Answer C is wrong. Valve lifters should be replaced if the bottom is pitted.
Answer D is wrong. Valve lifters should be replaced if the bottom is flat.

Question #79
Answer A is wrong.
Answer B is wrong.
Answer C is correct. Both a plugged transfer hose and a cracked filler neck soldered joint could prevent coolant from returning to the radiator when the engine cools.
Answer D is wrong.

Question #80
Answer A is wrong. Worn piston rings are likely to cause this condition.
Answer B is wrong. An obstructed PCV vacuum hose is likely to cause this condition.
Answer C is correct.
Answer D is wrong. A clogged PCV valve is likely to cause this condition.

Question #81
Answer A is correct. A cracked cylinder block would be unlikely to cause a hard start condition.
Answer B is wrong.
Answer C is wrong.
Answer D is wrong.

Question #82
Answer A is wrong.
Answer B is wrong.
Answer C is correct. On an engine equipped with electronic fuel injection, a loose intake manifold may cause both engine noise and poor vehicle driveability.
Answer D is wrong.

Question #83
Answer A is wrong. The thermostat should be fully open long before the water starts to boil.
Answer B is wrong. The thermostat should start to open when the water temperature reaches its rated temperature.
Answer C is wrong. Both Technicians A and B are wrong.
Answer D is correct.

Question #84
Answer A is wrong. Loose pistons may cause a rapping noise while accelerating.
Answer B is wrong. Worn cylinders may cause a rapping noise while accelerating.
Answer C is wrong. Worn main bearings may cause a thumping sound when starting.
Answer D is correct.

Question #85
Answer A is wrong. A belt tension gauge is used to check tension on standard V-belts.
Answer B is wrong. Measuring belt deflection on a V-ribbed belt is not the best method.
Answer C is wrong. A squealing noise at idle would indicate a loose or worn belt.
Answer D is correct.

Question #86
Answer A is wrong. Torque-to-yield bolts do stretch permanently when they are installed.
Answer B is wrong. Torque-to-yield bolts must be tightened in sequence.
Answer C is correct.
Answer D is wrong. Torque-to-yield bolts are usually tightened in a two-step procedure.

Question #87
Answer A is wrong.
Answer B is wrong.
Answer C is correct. Two common engine oil cooler mounting locations are: inside a radiator tank and ahead of the radiator support.
Answer D is wrong.

Question #88
Answer A is wrong. An anti-collapse spring is found in the lower, not the upper radiator hose.
Answer B is wrong. Low coolant level will not cause this noise.
Answer C is correct.
Answer D is wrong. Contact with power steering fluid causes a hose to become soft and gummy.

Question #89
Answer A is wrong.
Answer B is wrong.
Answer C is correct. Air trapped in the cooling system can cause engine overheating and a cracked cylinder head. Also, on some engines, trapped air can be released by unscrewing the coolant temperature sender.
Answer D is wrong.

Question #90
Answer A is correct.
Answer B is wrong. A ruler would not be used to take the measurement.
Answer C is wrong. Dial calipers are not the correct tool to measure the Plastigage width.
Answer D is wrong. A micrometer could not be used for this measurement.

Question #91
Answer A is wrong.
Answer B is correct. The reading displayed by the dial indicator must be divided by 2 to obtain stem-to-guide clearance.
Answer C is wrong.
Answer D is wrong.

Question #92
Answer A is wrong. The customer should be questioned, if possible.
Answer B is wrong. The customer complaint should always be eliminated.
Answer C is correct.
Answer D is wrong. A road test should always be performed to verify the repair.

Question #93
Answer A is wrong. A faulty ignition system is likely to cause a weak cylinder.
Answer B is wrong. A burned exhaust valve is likely to cause a weak cylinder.
Answer C is correct.
Answer D is wrong. A leaking intake manifold is likely to cause a weak cylinder.

Question #94
Answer A is wrong. Valve rotators cannot be disassembled.
Answer B is correct.
Answer C is wrong. Only Technician B is right.
Answer D is wrong. Technician B is right.

Question #95
Answer A is wrong. A press-fit harmonic balancer should be removed using a special balancer remove/install tool.
Answer B is correct.
Answer C is wrong. Only Technician B is right.
Answer D is wrong. Technician B is right.

Question #96
Answer A is wrong. The measurement gives no indication of journal condition.
Answer B is correct.
Answer C is wrong. The measurement would not indicate the camshaft lift.
Answer D is wrong. The measurement gives no indication of bearing clearance.

Question #97
Answer A is wrong. A lean fuel mixture would not cause a sulfur smell.
Answer B is wrong. Coolant leaking into the combustion chambers would cause a gray exhaust color.
Answer C is correct.
Answer D is wrong. A vacuum leak would cause a rough idle that would decrease as engine speed increases.

Question #98
Answer A is wrong.
Answer B is wrong.
Answer C is correct. In many cases, valve seat inserts are removable. Also, a newly installed valve seat insert should be staked in place.
Answer D is wrong.

Question #99
Answer A is wrong. Burned or leaking valves cause a fluctuation between 12 and 18 in. Hg (62 and 41kPa absolute).
Answer B is correct.
Answer C is wrong. Weak valve springs cause a fluctuation between 10 and 25 in. Hg (69 and 17 kPa absolute).
Answer D is wrong. A leaking head gasket would cause a fluctuation between 7 and 20 in. Hg (79 and 35 kPa absolute).

Question #100
Answer A is correct. Gear thickness and backlash specifications are rarely provided by manufacturers. The gears and housing should be inspected for scoring and other damage.
Answer B is wrong.
Answer C is wrong.
Answer D is wrong.

Question #101
Answer A is correct.
Answer B is wrong. A leaking exhaust valve would cause air to escape through the tailpipe.
Answer C is wrong. A leaking head gasket or a cracked head would cause air to escape from the radiator filler neck.
Answer D is wrong. Worn piston rings would cause air to escape from the opening in the rocker arm cover.

Question #102
Answer A is wrong. The freezing point of a coolant/water mixture starts to *increase* when coolant is more than 70 percent of the mixture.
Answer B is correct.
Answer C is wrong. Only Technician B is right.
Answer D is wrong. Technician B is right.

Question #103
Answer A is wrong.
Answer B is wrong.
Answer C is correct. After valve seats are resurfaced, the valve faces should be coated with blue dye and the valves should be installed in the head. If blue dye does not transfer to the head all the way around the seat, replace the valve because it is bent or improperly ground.
Answer D is wrong.

Question #104
Answer A is wrong. 5 inches (127 mm) is too close.
Answer B is correct.
Answer C is wrong. 7 inches (179 mm) is too far away.
Answer D is wrong. 8 inches (203 mm) is too far away.

Question #105
Answer A is wrong. This tool cannot check combustion chamber cubic centimeters (cc).
Answer B is correct.
Answer C is wrong. Only Technician B is right.
Answer D is wrong. Technician B is right.

Question #106
Answer A is wrong. Worn intake valves would cause air leaks at the throttle body or carburetor.
Answer B is wrong. Worn exhaust valves would cause air leaks at the tailpipe.
Answer C is wrong. A broken PCV valve would not cause air to leak.
Answer D is correct.

Question #107
Answer A is wrong.
Answer B is wrong.
Answer C is wrong.
Answer D is correct. Three drops every 100 feet would result in the loss of 3 quarts (2.8 liters) every 1,000 miles (1,609 km).

Question #108
Answer A is wrong. Tensioner length should be measured, not the chain tension.
Answer B is wrong. Oil feed pressure would not indicate chain wear.
Answer C is correct.
Answer D is wrong. An engine equipped with a hydraulic chain tensioner should not be rotated backwards.

Question #109
Answer A is wrong. On OHC engines, the timing belt or chain does not have to be removed from the block before the head can be removed. In many cases, the camshaft sprocket is disconnected from the cam and the cylinder head is lifted from the engine, leaving the belt or chain in place.
Answer B is correct.
Answer C is wrong. Only Technician B is right.
Answer D is wrong. Technician B is right.

Question #110
Answer A is correct.
Answer B is wrong. Removing the spark plugs is not the first step that should be taken.
Answer C is wrong. The ignition system should be disabled before the engine is rotated by hand.
Answer D is wrong. The spark plugs must be removed before you can check for oil or coolant flow in the spark plug holes.

Question #111
Answer A is wrong.
Answer B is wrong.
Answer C is correct. Both stuck piston rings and a plugged oil drain passage in the cylinder head may allow excessive oil to enter the cylinders. This oil would produce blue/gray smoke when it burned.
Answer D is wrong.

Question #112
Answer A is wrong. The seals are installed after the valves are installed.
Answer B is correct.
Answer C is wrong. Only Technician B is right.
Answer D is wrong. Technician B is right.

Question #113
Answer A is wrong. The thread created will be the same size as the original thread.
Answer B is wrong. The thread created will not be smaller than the original thread.
Answer C is wrong. The thread created will not be larger than the original thread.
Answer D is correct.

Question #114
Answer A is correct.
Answer B is wrong. A clogged radiator would not cause this.
Answer C is wrong. Only Technician A is right.
Answer D is wrong. Technician A is right.

Question #115
Answer A is wrong. The wastegate diaphragm cannot be damaged by too much boost.
Answer B is wrong. Too much boost cannot damage compressor wheel blades.
Answer C is wrong. Both Technicians A and B are wrong.
Answer D is correct.

Question #116
Answer A is wrong. Coating both sides of the rubber seals increases the likelihood that the seals will be squeezed out of place when the intake manifold bolts are tightened.
Answer B is correct.
Answer C is wrong. Only Technician B is right.
Answer D is wrong. Technician B is right.

Question #117
Answer A is wrong. Excessive runout would cause binding.
Answer B is wrong. Improperly installed bearings would cause binding.
Answer C is wrong. Bore misalignment would cause binding.
Answer D is correct.

Question #118
Answer A is wrong. Fluorescent light will not cause the dye to glow.
Answer B is correct.
Answer C is wrong. A strobe light will not cause the dye to glow.
Answer D is wrong. Infrared light will not cause the dye to glow.

Question #119
Answer A is wrong. The technician should verify that the woodruff key is in place.
Answer B is wrong. The technician should verify that the oil slinger is in place.
Answer C is correct.
Answer D is wrong. The technician should verify that the oil seal has been lubricated.

Question #120
Answer A is correct.
Answer B is wrong. If the starter ring gear on an automatic transmission flywheel is damaged, the flywheel/ring gear assembly must usually be replaced.
Answer C is wrong. Only Technician A is right.
Answer D is wrong. Technician A is right.

Question #121
Answer A is correct.
Answer B is wrong. Valve stems and valve guides should be lubricated with engine oil prior to assembly.
Answer C is wrong. Only Technician A is right.
Answer D is wrong. Technician A is right.

Question #122
Answer A is wrong.
Answer B is wrong.
Answer C is wrong.
Answer D is correct. A stuck shut heat riser valve will force exhaust gases through the underside of the intake manifold at all times, resulting in manifold overheating and possibly cracking.

Question #123
Answer A is correct.
Answer B is wrong. Camshaft runout cannot be measured with the camshaft still mounted in the engine.
Answer C is wrong. Only Technician A is right.
Answer D is wrong. Technician A is right.

Question #124
Answer A is wrong. Using a screwdriver to pierce and pull on the hose may damage the heater core.
Answer B is wrong. Use of a pliers and excessive force may damage the heater core.
Answer C is wrong. Use of a prying tool and excessive force may damage the heater core.
Answer D is correct.

Glossary

Air conditioning The process of adjusting or regulating, by heating or cooling, the quality, temperature, and humidity of air.

Aluminum A nonferrous metal that is light in weight yet can be stronger than steel when mixed with the proper alloys. It is easily cast and machined.

Balance shaft A shaft with counterweights designed to prevent vibration of rotating parts.

Battery A device for storing electrical energy in chemical form.

Belt A device used to drive the water pump and other accessory power-driven devices.

Block deck The flat surface of the main casting of an engine on which the head attaches.

Cam An abbreviation for camshaft; a device having lobes, driven by the crankshaft via gears, a chain, or a belt that opens and closes the intake and exhaust valves.

Cam follower A term used for valve lifter; a hydraulic or mechanical device, in the valve train, that rides on the camshaft lobe to lift the valve off its seat.

Camshaft journal That part of the camshaft that turns in a bearing.

Cast iron A term used for various cast ferrous alloys containing at least 2% carbon. It is used for many different parts on vehicles.

Catalytic converter An automotive exhaust system component, made of stainless steel, containing a catalyst that reduces hydrocarbons, carbon monoxide, and nitrogen oxides present in the engine exhaust gases.

Coil A term used to describe a spring or an electrical device using many turns (coils) of wire such as springs, ignition coils, solenoids, and relays.

Connecting rod bearing The bearing of a connecting rod that rotates on the crankshaft.

Cooling system The system that circulates coolant through the engine to dissipate its heat.

Counterbalance shafts One or more rotating shafts found on some engines to counteract the natural vibrations of other rotating parts, such as the crankshaft, in that engine. This balance, or counterbalance, shaft usually turns at twice the crankshaft speed and must be timed properly.

Crankshaft The revolving part of a unit that has the function of delivering power or work from the reciprocating motion. Engines have crankshafts, as do air conditioning compressors and air compressors.

Crankshaft sensor An electronic device used to send crankshaft rotating information to the computer.

Cross firing A condition whereby spark plugs fire out of turn, usually caused by poor spark plug wire insulation.

Cylinder head That part of the engine that covers the cylinders and pistons.

Cylinder wall A term used for cylinder bore; the inside diameter of a cylinder.

Distributor A device used on many engines to direct high-voltage electrical energy from the coil to the spark plugs.

Elongated Not round; egg-shaped.

End play A term used to describe a spacing or clearance involved with a moving part. Crankshafts and camshafts require end play measurements and must be set to manufacturer's specifications during engine assembly.

Engine oil A lubricant formulated for use in an engine.

Engine oil cooler A device used on some high performance engines, police packages, taxis, trucks, turbo-equipped engines, and diesel engines to prevent the engine oil from overheating. These work by using a heat exchanger exposed to air flow or engine coolant.

Exhaust The burned and unburned gases that remain after combustion.

Exhaust pipe A pipe that connects the exhaust manifold to the muffler or catalytic converter. It is made of heavier material than tailpipes, and sometimes is double layered.

Face-to-seat As in valve face-to-valve seat contact, this refers to the actual sealing area of the valve and seat. It must be the size and angle specified by the engine manufacturer to properly seal the combustion chamber and have a long service life.

Firing order The order in which the engine cylinders fire and deliver power.

Flywheel A round, heavy metal plate attached to the crankshaft of an engine that helps smooth out power strokes and gives the rotating crankshaft momentum to smoothly get to the next power stroke. The clutch and pressure plate help the flywheel transmit power to the drive wheels.

Frozen A mechanical problem developed from a lack of oil or broken internal parts that prevents motion, such as an engine rotating.

Garter spring A small spring placed behind the lip of a lip seal to maintain contact with the rotating part.

Gap A space between two adjacent parts or surfaces.

Head That part of an engine that covers the top of the cylinders and pistons.

Heat shield Devices used many places on today's vehicles. One such place is between the starter solenoid and the heat of the engine and exhaust manifold. Another is between electrical wiring or spark plug wiring and any high heat source. Heat shields are also used between catalytic converters and the passenger compartment of the vehicle.

Hone To use abrasive materials to remove material from a surface, or just to change the surface smoothness so the parts involved will work better.

Idler pulley A pulley that is used to adjust the belts on a belt-driven system.

Input A term used for the signals sent to different electronic modules about operating conditions of systems involved.

Intake ductwork All of the connecting ductwork from the throttle body out involved with getting air into the engine. This ductwork can collapse and restrict air flow into the engine. If this ductwork cracks or has loose connections, the engine could run poorly, because unmeasured air would be entering the engine. If the fresh air end of the ductwork is exposed to an area that could suck standing water, snow, or rain water into the engine while driving, serious engine damage would likely result.

Jumped As in jumped timing chain or timing belt, resulting in a condition where the valve timing is no longer correct, and the engine will run poorly or might not start.

Keepers Key-like tapered metal locking devices used to hold valve retainers in place.

Lash The clearance between two parts.

Lash adjuster A device much like hydraulic lifters that is usually found on overhead cam engines. Its job is to maintain zero lash between the cam follower and the tip of the valve.

Line boring A machining process that ensures multiple holes that are bored in a cylinder head or block are in line or true. This allows the camshaft or crankshaft installed in these locations to turn freely and function properly. As in line bore alignment of camshaft bearing or main bearing bores, a special boring bar is used that removes metal from all cam or main bearing bores at the same time and in a straight and true line.

Main bearing cap The structural device that holds the crankshaft in place in an engine block.

Muffler A device in the exhaust system used to reduce noise.

No-crank A condition like a frozen engine, a defective starter, or an electrical problem that prevents the engine from rotating when normal attempts are made to start the engine.

No-start A condition where the engine turns over normally but does not start. This could be caused by mechanical problems, fuel system problems, electrical problems, or electronic engine control problems.

Oil cooler A heat exchanger used to cool transmission or engine oil. Also may be used to cool power steering or other fluids.

Oil filter A device used to remove impurities, such as abrasive particles, from oil.

Oil pan A removable part of the engine assembly that contains the oil supply.

Out of square As in a valve spring that is not within specification when checked vertically on a flat surface and measured at a 90-degree angle. If installed in an engine, such a spring would cause side pressures on the valve and damage the valve and valve guide.

Overhead cam A camshaft that is mounted in the cylinder head.

Oxygen sensor An electronic device found in the exhaust system that measures the amount of oxygen in the exhaust stream.

Piston pin A precision ground pin, usually hollow, used to attach the connecting rod to the piston. The piston pin can be held in place by a press fit, snaprings, or bolts.

Pressure cap A cap placed on the radiator to allow regulated, above-atmospheric pressure in the cooling system.

Primed Ready; prepared.

Pulley A wheel-shaped device used in a belt-drive system to drive accessory equipment.

Ram air Air forced through the radiator, condenser, and across the engine by the forward movement of the vehicle.

Reluctor A gear-like part of an electronic ignition system. It could have the same number of teeth (or gaps) as cylinders of the engine, or it could have half the number of teeth as the engine has cylinders. As a tooth (or gap) passes by a pickup coil or crankshaft sensor, the magnetic field changes, and a trigger signal is sent to an electronic control module.

Reluctor ring A gear-like part of the electronic ignition system.

Rotators As in valve rotators—devices that cause the valves in the cylinder head to rotate slightly each time the valve closes. This helps keep the valve seat and valve face clean. Because a different part of the valve face contacts the valve seat each time it closes, the valve also operates at a cooler temperature. This greatly extends valve and seat life.

Rotor A part of the ignition distributor that rotates inside the cap and transfers ignition coil secondary electrical energy from the center tower to the individual spark plug wires.

RTV An abbreviation for Room Temperature Vulcanizing; the trade name for a rubber-like sealing compound.

Select fit As in the main bearings of an engine. Many of today's engines use select fit main bearings. This means the main bearings are no longer all one standard size, or all 0.010 inch (0.254 mm) undersized. The manufacturer mixes and matches bearing halves of small increments of as little as 0.003 inch (0.076 mm) to ensure better oil and noise control at the crankshaft area.

Selective thrust washer A washer or spacer that is furnished in different sizes to facilitate end play and preload settings.

Skirt A term used to describe the lower part of an engine piston. The skirt contacts the cylinder wall, helps the piston travel in a straight line, and prevents piston slap.

Solenoid An electromechanical device used for a push-pull operation.

Span A term used for the length or space between two parts; as in the span between the air conditioner compressor belt pulley and the alternator pulley is too great. This condition can cause the belts to fly off at high engine speeds or during rapid acceleration.

Spark plug A component of the ignition system that delivers the high-voltage spark to the combustion chamber.

Starter drive The part of the starter motor that engages the ring gear or the flywheel, flex plate, or torque converter, and rotates the engine on start-up.

Starter motor The small electric motor that is used to crank (start) an engine.

Starter solenoid That part that causes the starter drive to engage the flywheel when starting an engine.

Tailpipe The pipe from the muffler or catalytic converter that carries the exhaust gases away from the passenger compartment.

Temperature sensor A term used for various temperature sensing switches or variable resistors on today's vehicles. They can be used to turn cooling fans on and off, to operate dash gauges, to control air conditioners, and to furnish inputs to engine and transmission control computers.

Tensioner A device used with timing belts and timing chains that maintains a constant pressure on the belt or chain to minimize wear and noise, and to take up slack or lost motion. They may be fixed and require periodic adjustment, or they may be automatically adjusted by spring tension or engine oil pressure. Tensioners are also used on accessory drive belts.

Timing belt The belt through which the crankshaft drives the camshaft(s) in an overhead cam engine.

Torque wrench A specially designed tightening device that indicates the amount of torque being exerted on a fastener to enable threaded parts to be tightened to a specified amount.

Torque-to-yield A term used to describe a common method of tightening fasteners on many of today's engines. The procedure is to tighten the fasteners to a fairly low pounds-feet value, then to turn each fastener (in proper tightening sequence) a specified number of degrees. This process is usually used on head bolts, main bearing bolts, and rod bearing bolts. In most cases, new fasteners are required for every reassembly.

True A term used in the automotive industry to signify a part or system is correct and is within specifications. As in the cylinder head gasket sealing surface is true; i.e., it is not warped, scratched, broken, or otherwise damaged.

Valve float A condition that occurs when the valve spring is not capable of closing the valve quickly enough. This usually happens at higher engine speeds, and is aggravated by the valve spring losing some of its tension. Improper sealing of the combustion chamber will occur and the engine will run poorly.

Valve lifter A hydraulic or mechanical device in the valve train that rides on the camshaft lobe to lift the valve off its seat.

Valve rotator A device that rotates the valve while the engine is running.

Valve spring retainer A device on the valve stem that holds the spring in place.

Valve train The parts making up the valve assembly and its operating mechanism.

Warpage As in cylinder head gasket sealing area no longer being straight and true. This condition will require machine shop work on the cylinder head or a different cylinder head that is not warped.

Water pump A mechanical device used to circulate coolant through the cooling system.

Witness marks Lines scribed on adjacent surfaces of mating parts, before disassembly, to ensure proper alignment when reassembled.

Resource List

Hollembeak, Barry. *Classroom Manual for Automotive Engine Repair and Rebuilding.* Today's Technician. Albany, New York: Delmar Publishers, 1997.

Hollembeak, Barry. *Shop Manual for Automotive Engine Repair and Rebuilding.* Today's Technician. Albany, New York: Delmar Publishers, 1997.